AI ARTISTRY

A Beginner's Guide to Generative AI

By
Chris Elliott

AI ARTISTRY

A Beginner's Guide to Generative AI

CONTENTS

INTRODUCTION

Welcome to the exciting world of generative artificial intelligence, a realm where creativity and technology intersect in fascinating and unexpected ways. This book aims to unravel the intricacies of generative AI, making it accessible even to those who are new to the concept. Whether you're an eager beginner or an enthusiast with a curiosity for tech-savvy creativity, this book is designed to be your comprehensive guide, leading you through the basics and diving into the endless possibilities of generative AI.

Generative AI is not just a technological marvel; it's a revolutionary tool that has the power to augment human creativity and transform various forms of digital art. At its core, it involves machines creating data, art, music, or even text, not by manually coding each step but by learning patterns and generating new outputs from them. Imagine a painter's brush subtly guided by an algorithm or a melody composed by a neural network. The creative spectacle made possible by generative AI is truly boundless, limited only by the imagination of its human collaborators.

Technology often seems daunting due to its complexity, but generative AI can be fascinating without being intimidating. Our journey begins with a straightforward and accessible introduction to fundamental principles and concepts. We'll explore the magic behind how machines learn, create, and inspire. As we delve deeper, we'll unlock the secrets behind buzzwords like neural networks, GANs (Generative Adversarial Networks), and VAEs (Variational Autoencoders).

One might wonder, why now? Why has generative AI become such a focal point in contemporary discussions around technology and creativity? It's the perfect storm of years of refined machine learning techniques meeting exponentially improving computational power. What was once the domain of highly specialized researchers is now accessible to enthusists and artists around the globe. Machines have become collaborative partners in creative endeavors, democratizing artistic production in ways previously unimaginable.

Our exploration begins with a historical context, understanding the evolution of generative AI from its theoretical underpinnings to its current practical applications. This historical perspective not only sheds light on how far we've come but also sets the stage for appreciating the nuances of modern AI techniques and tools.

Without giving too much away, let's touch briefly on what you'll encounter in the following chapters. Starting with the fundamental principles of machine learning, we'll tackle the basics before diving into the more specialized architecture like neural networks. These networks are the backbone of generative AI, enabling models to mimic human-like learning and creativity. Special focus will be placed on Generative Adversarial Networks (GANs) and Variational Autoencoders (VAEs), the linchpins of generative art.

As you grow more comfortable with the underlying technology, we'll introduce you to the various tools and software used to create generative art. From setting up your workspace to implementing your first model, comprehensive guides will assist you every step of the way. Given the importance of data, there will also be extensive discussion on data collection and preparation. You'll learn how to ethically source and manage data, ensuring your creations respect legal and societal norms.

The heart of generative AI lies in its applications. Imagine a world where machines generate stunning visual art, compose music that pulls at your heartstrings, or even create immersive environments for video games and virtual reality. The chapters on creating visual art, generating music, and building interactive installations will showcase techniques and offer practical insights into bringing your own AI-powered creations to life.

What about the societal impact? It's crucial to address the broader implications of this technology. Ethical considerations, questions of ownership, and the evolving landscape of AI art criticism are discussed in dedicated chapters. As generative AI continues to blur the lines between human and machine creativity, understanding these issues becomes paramount.

Real-world applications and inspiring case studies will be peppered throughout the book, highlighting notable projects and artists who have pioneered this space. Their stories offer both inspiration and invaluable lessons, illustrating what's possible when human creativity is coupled with machine intelligence.

For those ready to dive into hands-on projects, practical guides will provide step-by-step instructions to kickstart your AI art journey. You'll find resources, templates, and troubleshooting tips to keep you moving forward. Additionally, we'll discuss how to maintain and update your AI models, ensuring your work stays relevant and continues to evolve.

As if that weren't enough, we'll explore the world of monetizing AI art, from licensing and sales to innovative methods like crowdfunding and sponsorship. Our goal is to empower you to not only create but also share and sustain your work. Finally, we'll cast our eyes to the future, predicting emerging trends and technologies that might further revolutionize the field of generative AI.

In conclusion, this book is more than a technical manual; it's an invitation to explore new frontiers of creativity. Generative AI has bridged the gap between art and science, and its potential continues to grow. By the end of this journey, you'll have both the knowledge and the inspiration to contribute to this rapidly evolving field, creating works that are uniquely your own.

So prepare to immerse yourself in the dynamic, inspiring, and sometimes bewildering world of generative AI. Your journey into the confluence of technology and creativity starts here, and the only limit is your imagination.

CHAPTER 1:
WHAT IS GENERATIVE AI?

Generative AI represents a burgeoning frontier in the field of artificial intelligence, shifting from merely observing and analyzing the world to creating new, imaginative outputs. At its core, generative AI involves algorithms that can produce original content—from art and music to prose and code—by learning patterns from existing data. It harnesses sophisticated models, like neural networks, to generate outputs that often mirror human creativity. The evolution of these technologies promises not just to enhance artistic expression but to redefine it, offering novice enthusiasts a gateway into the captivating interplay between machine learning and creativity. As we delve into the history, definition, and implications of generative AI, we'll uncover the transformative potential that lies at the intersection of innovation and artistic endeavor.

Defining Generative AI

At its core, generative AI is a subfield of artificial intelligence focused on creating models that can generate outputs indistinguishable from those created by humans. But what exactly does that mean? Picture an AI that can compose music, design intricate visuals, or even pen down a short story. Generative AI does all this by learning patterns from existing data and utilizing that knowledge to create new, original content. It's fascinating and, for those beginning to explore this area, an eye-opening gateway to the possibilities of machine creativity.

Generative AI operates by training on large datasets, which serve as a kind of creative inspiration. By analyzing parameters, styles, and essentials present in the data, these AI models learn the nuanced details that make a piece of art or a piece of writing unique. Once trained, these models can then produce new content in a similar style, often with a level of creativity that can surprise even the most seasoned artists or writers. This approach is fundamentally different from traditional AI models that focus on tasks like image recognition or language translation, which are centered more on classification and prediction rather than creation.

Imagine having a virtual artist who has studied thousands of classic paintings. This virtual artist can then produce a brand-new painting that captures the essence of any selected art period. That's the kind of ability generative AI brings to the table. It's like having a digital assistant that not only understands past works but can also contribute to future creative endeavors.

One of the central methods used in generative AI is known as Generative Adversarial Networks (GANs). GANs consist of two neural networks: a generator and a discriminator. The generator creates content while the discriminator evaluates it. Through this adversarial process, the generator continually improves its creations, making them increasingly realistic. This game of one-upmanship between the two networks results in outputs that can be astonishing in their fidelity to human-created works. However, we'll dive deeper into GANs in a later chapter.

Another significant approach under the generative AI umbrella is Variational Autoencoders (VAEs). VAEs work by encoding data into a latent space and then decoding it back to recreate the original data. By tweaking this latent space during the process, VAEs can generate entirely new instances of data that echo the characteristics of the original dataset. VAEs offer a structured way to understand the data's underlying distribution and generate new data points accordingly. They balance between capturing a dataset's variance and generating coherent, meaningful outputs.

The broader landscape of generative models doesn't stop with GANs and VAEs. Other techniques include autoregressive models and flow-based models, each with its unique strategies for creating new and diverse content. Autoregressive models generate data step-by-step, with each step conditioned on the previous ones, producing highly coherent sequences such as text or music. Flow-based models, on the other hand, employ invertible transformations to map data to a latent space, providing a different route to generation that's mathematically elegant and efficient.

What sets generative AI apart is its incredible versatility. While traditional AI is often bounded by predefined rules and outcomes, generative AI can explore the uncharted territories of creativity. In essence, it's not just solving problems; it's creating new things we didn't know we needed. Whether it's generating synthetic yet realistic photographs, crafting new musical compositions, or writing articles, generative AI is pushing the envelope of what machines can achieve.

The future of generative AI holds immense promise. As these models continue to evolve, they will likely integrate more seamlessly with various creative and professional fields. From enhancing the capacity for personal expression to revolutionizing industries like design, entertainment, and education, the applications are manifold. Consider how advertising agencies can use generative AI to produce bespoke campaigns quickly, or how educators can deploy AI-generated content tailored to individual learning needs. The possibilities are expansive and thrilling to contemplate.

Generative AI's roots lie in machine learning and neural networks, where the focus shifted from merely classifying data to creating it. This transition signifies a shift from understanding the world solely through analysis to also contributing to it through synthesis. This paradigm shift in AI's capabilities allows for a deeper interaction between humans and machines, fostering collaboration where both parties can contribute their unique strengths.

Despite its numerous advantages, generative AI also poses challenges. Issues around copyright and the ethical use of generated content are significant. The line between what is human-made and what is AI-generated can blur, raising questions about authenticity and ownership. For instance, if an AI composes a piece of music, who owns the rights to that music? The developer, the user, or the AI itself? These are questions that society will need to address as the technology becomes more widespread.

Additionally, the quality of generated content is highly dependent on the data fed into the models. Biases present in training data can lead to biased outputs, a problem that can pervade art, text, and other generated content. As the creators and curators of this technology, we must be vigilant about the kind of data we use and the way it influences our models.

Generative AI also opens avenues for personalized experiences. Imagine tailoring a book's storyline to suit a reader's preferences or customizing art according to an individual's taste. These personalized touches could redefine consumer experiences, making interactions more meaningful and unique. Such dynamic capabilities offer a more inclusive approach to creativity, accommodating a wider range of tastes and preferences.

Ultimately, generative AI stands at the confluence of technology and creativity. Its potential to transform industries and redefine our understanding of creativity is monumental. For beginners and enthusiasts, understanding the fundamentals of generative AI is the first step toward tapping into this transformative power. The journey ahead is not just about learning how these models work; it's about exploring how they can work for you, empowering your ideas and amplifying your creativity.

History and Evolution

Generative AI has roots that run deep into the vast tapestry of artificial intelligence research. Going back to the mid-20th century, the journey began with the advent of the earliest computer algorithms. These laid the groundwork for what would eventually become an intricate and powerful domain of machine learning. Understanding the history of generative AI gives us an appreciation for how far we have come and where we might be headed.

In its infancy, artificial intelligence research concentrated on symbolic AI— a stark contrast to what we now recognize as generative models. Symbolic AI was rule-based and required explicit instructions for problem-solving. These early systems were constrained by their inability to adapt and learn from data in the way that modern generative AI can. The development of neural networks in the 1950s and 1960s marked a critical juncture. Early pioneers like Frank Rosenblatt and his Perceptron demonstrated that a machine could learn from data, albeit in a limited capacity.

However, the capabilities of neural networks at that time were rudimentary. For many years, progress stagnated due to various limitations, including computational power and lack of data. This period, known as the "AI Winter," saw diminished interest and funding in the field. It wasn't until the late 1980s and early 1990s that the renaissance for AI began, driven by advancements in both hardware and theoretical foundations. Researchers discovered Backpropagation, a method for training multi-layered neural networks, which reignited optimism and innovation in the field.

Fast forward to the mid-2000s, and we start to see the emergence of Generative Adversarial Networks (GANs), a revolutionary concept introduced by Ian Goodfellow and his colleagues in 2014. GANs fundamentally transformed the landscape of generative AI by implementing a dual-model approach: a generator and a discriminator. This setup allowed the generator to create data resembling the training data, while the discriminator distinguished between real and generated data. The two models trained in conjunction, continually improving each other's performance.

The introduction of GANs opened the floodgates for a plethora of generative models spanning various applications. Researchers quickly started leveraging this architecture to produce highly realistic images, lifelike videos, and even coherent text. The creative potential of generative AI seemed almost limitless.

Another significant milestone in the history of generative AI was the development of Variational Autoencoders (VAEs). Unlike GANs, VAEs focus on learning embeddings of the input data that allow for both reconstruction of the input and generation of new samples. Introduced by Kingma and Welling in 2014, VAEs have been critical in generating data that follows the distribution of the training set, thus offering a different set of capabilities and use cases compared to GANs.

The evolution of generative AI also owes much to the advent of large-scale datasets and rapid advancements in computational power. High-performance GPUs and TPUs enabled the training of increasingly complex models on massive datasets. The advent of big data provided the necessary fuel for these models to learn intricate patterns and generate high-quality outputs. With the rise of cloud computing, researchers had more access than ever to computational resources, democratizing the field and enabling a broader community to contribute to advancements.

In parallel with hardware advancements, the evolution of software tools and frameworks accelerated progress. Libraries like TensorFlow,

AI Artistry

PyTorch, and Keras simplified the process of building and training generative models, making it more accessible to researchers and enthusiasts alike. These open-source platforms became the backbone of countless groundbreaking projects, allowing ideas to move from concept to implementation at unprecedented speeds.

As we ventured into the 2020s, the capabilities of generative AI continued to expand. Models became increasingly sophisticated, incorporating techniques such as attention mechanisms and transformers. These innovations gave rise to highly capable language models like GPT-3, capable of generating human-like text with impressive coherence and context-awareness. The impact of such models has been profound, influencing industries ranging from marketing to entertainment.

The journey of generative AI doesn't stop here. The relentless pace of research and development suggests that we will continue to witness unprecedented advancements. Hybrid models that combine the strengths of various architectures are already beginning to take shape. Additionally, the ethical considerations and societal implications of generative AI are becoming front-and-center, prompting discussions and research on making generative AI more responsible and equitable.

The Historical and evolutionary lens of generative AI serves not just as a chronology but as a narrative of human ingenuity and relentless pursuit of creativity and intelligence. Every milestone achieved has paved the way for even more incredible breakthroughs, making it an exhilarating field to be a part of. As we look to the future, the possibilities seem boundless, and the journey of generative AI continues to inspire and captivate the imagination of enthusiasts and professionals alike.

11

CHAPTER 2:
BASICS OF MACHINE LEARNING

Machine learning lies at the heart of generative AI, serving as the engine that powers the creation of new, often startlingly innovative content. In its most basic form, machine learning is about teaching computers to learn from data and make decisions or predictions based on that information. Imagine having a huge data set of images, texts, or sounds, and then training an algorithm to understand patterns and generate something entirely new from these examples. This process involves key concepts like training data, algorithms, and model evaluation, each essential to creating systems capable of performing complex tasks with impressive accuracy. As you start to grasp these fundamentals, it becomes clear how machine learning transforms from a theoretical framework into a tool of profound creative potential. This chapter will guide you through these basics, setting the foundation you'll need to explore the fascinating world of generative AI.

Understanding Machine Learning

Machine learning acts as the backbone of generative AI, enabling machines to learn from data and make decisions with minimal human intervention. But before diving deep into applications, it's essential to understand what machine learning really entails. Imagine teaching a child to recognize different animals. Over time, with enough examples and corrections, the child gets better at identifying them. Similarly, machine learning algorithms improve their performance by being exposed to more data.

At its core, machine learning involves feeding massive quantities of data into algorithms, allowing those algorithms to detect patterns and relationships within the data. These patterns are then used to make predictions or decisions without being explicitly programmed for every possible scenario. For instance, a machine learning model trained on thousands of cat images can distinguish new cat images from dog images it hasn't seen before.

Machine learning is often divided into three primary types: supervised learning, unsupervised learning, and reinforcement learning. Supervised learning involves training an algorithm on a labeled dataset. For example, if you want a model to recognize handwritten digits, you would supply it with a dataset of images of handwritten digits, each labeled with the correct number. The model learns by comparing its predictions to the labels and adjusting accordingly.

Unsupervised learning, on the other hand, deals with unlabeled data. Here, the algorithm tries to identify inherent patterns and structures within the data without any specific instructions. This approach is often used in clustering tasks where the goal is to group data points that are similar to one another. For example, customer segmentation in marketing can benefit from unsupervised learning to identify distinct groups of customers based on their purchasing behavior.

Lastly, reinforcement learning is inspired by behavioral psychology and involves learning through trial and error. An agent, placed in an environment, learns to perform tasks by receiving feedback in the form of rewards or penalties. Over time, it aims to maximize the cumulative reward. One of the most famous applications of reinforcement learning is in game-playing AI, like AlphaGo, which learned to play the game of Go by playing millions of matches against itself.

Central to all these learning paradigms are the data and the algorithms. Data is often considered the "fuel" for machine learning models. The quality, quantity, and relevancy of data directly impact the performance of the models. On the other hand, algorithms act as the "engines," processing data to extract useful patterns. Algorithms can range from simple, like linear regression, to complex, like deep neural networks, which mimic the structure of the human brain.

In addition to the types of learning, understanding key concepts like overfitting and underfitting is crucial for anyone delving into machine learning. Overfitting occurs when a model learns the training data too well, including the noise, and fails to generalize to new, unseen data. It's akin to a student who memorizes answers rather than understanding concepts. Underfitting, conversely, happens when a model is too simple to capture the underlying structure of the data, leading to poor performance on both training and test data.

Regularization techniques, cross-validation, and hyperparameter tuning are some methods used to strike a balance between underfitting and overfitting. Regularization adds a penalty to the model's complexity, discouraging it from fitting to noise. Cross-validation involves dividing the data into multiple subsets to ensure the model performs well on different data samples. Hyperparameter tuning involves finding the best parameters that control the learning process.

The choice of algorithm and model architecture also plays a pivotal role in machine learning. Linear models are straightforward and computationally efficient but may fall short for complex tasks. Non-linear models like decision trees and random forests can capture more intricate relationships but may require more computation. Neural networks, especially deep learning models with multiple layers, have shown extraordinary success in tasks involving images, audio, and text but come at the cost of increased computational resources and complexity in training.

Beyond the technical aspects, the success of machine learning projects often hinges on the iterative process of building, evaluating, and refining models. This workflow typically begins with data collection and preprocessing, which includes cleaning data, handling missing values, and engineering features. Next comes model training, where different models are trained and their performance evaluated using metrics like accuracy, precision, recall, and F1-score, depending on the task at hand. Finally, models are often deployed into production, where they are monitored and updated as new data becomes available.

It's not just about choosing the right algorithm or collecting more data; the art of machine learning lies in understanding the problem at hand and concocting an appropriate strategy to solve it. Whether it's predicting stock prices, diagnosing diseases, or creating generative art, machine learning offers a broad toolkit that, when wielded wisely, can achieve remarkable results.

Various tools and frameworks have democratized access to machine learning, making it easier for enthusiasts and beginners to get started. Popular tools like TensorFlow, PyTorch, and Scikit-learn provide robust libraries for building and training machine learning models. Additionally, platforms like Jupyter Notebook allow for interactive coding, which is especially useful for experimenting with different models and visualizing results.

Of course, machine learning doesn't exist in a vacuum. It builds upon statistical theories and computer science principles. Understanding concepts like probability distributions, optimization algorithms, and computational complexity can deepen one's comprehension and application of machine learning techniques. Furthermore, staying updated with the latest research and methodologies by following reputable journals, attending conferences, and participating in online communities can contribute significantly to one's growth in this field.

The realm of machine learning is expansive, and its influence is ever-growing. As we delve deeper into the intricacies and explore its myriad applications, machine learning holds the promise of transforming industries and pioneering new creative possibilities. From automating mundane tasks to pushing the boundaries of art and science, the journey into machine learning is as exhilarating as it is enlightening.

Thus, understanding machine learning is fundamental for anyone eager to harness the power of generative AI. It provides the essential knowledge and tools required to embark on a journey of innovation and creativity, pushing the limits of what machines can achieve alongside humans. With a firm grasp of these basics, you're well on your way to exploring and creating awe-inspiring works with generative AI.

Key Concepts and Terminologies

Understanding the world of machine learning requires familiarizing oneself with the various foundational terms and concepts. Machine learning, a subset of artificial intelligence, allows systems to learn from data, identify patterns, and make decisions without explicit programming. Central to this discipline are a host of key concepts and terminologies, each vital to grasping the broader picture and engaging creatively with generative AI.

Algorithm: An algorithm is a set of rules or instructions given to a machine to help it achieve a specific task. Think of it as a recipe; just as a recipe guides a chef to create a dish, an algorithm guides a computer to solve a problem or perform a computation. In machine learning, algorithms are used to extract patterns from data.

Training Data and Test Data: Two critical types of data used in machine learning are training data and test data. *Training data* is the dataset on which the model is trained—that is, from which the model learns the patterns and relationships. On the other hand, *test data* is a separate dataset used to evaluate the model's performance. The distinction ensures that the model can generalize its learning to new, unseen data.

Model: In machine learning, a model is a mathematical representation of a real-world process. It's constructed by training an algorithm on historical data and is used to make predictions or decisions. Models can vary in their complexity and the amount of data they require to be effective.

Features and Labels: Features are the input variables used to make predictions. For instance, in predicting house prices, features might include square footage, number of bedrooms, and neighborhood quality. The label, on the other hand, is the output variable that the model aims to predict—in this case, the house price.

Supervised Learning: Supervised learning is a type of machine learning task where the model is trained on labeled data. The training dataset includes both the input features and the correct output labels. The goal is for the model to learn the mapping from inputs to outputs and make accurate predictions on new, unseen data.

Unsupervised Learning: Unlike supervised learning, unsupervised learning involves training a model on data that does not include labels. The model attempts to discover intrinsic patterns in the data. A common application is clustering, where the model groups similar data points together.

Overfitting and Underfitting: Overfitting occurs when a model learns the noise in the training data too well, leading to excellent performance on the training data but poor generalization to new data. Underfitting happens when the model is too simple to capture the underlying trends in the data, resulting in poor performance on both training and test sets. Striking the right balance is crucial for effective machine learning.

Neural Networks: Neural networks are a class of models inspired by the human brain's structure and function. They consist of layers of interconnected nodes or neurons, each performing a simple computation. When combined, these neurons can model complex patterns in the data. Neural networks are particularly powerful for tasks like image and speech recognition.

Activation Function: Within a neural network, an activation function determines whether a neuron should be activated based on the weighted sum of its inputs. Common activation functions include sigmoid, tanh, and ReLU (Rectified Linear Unit). The choice of activation function can have a significant impact on the network's performance.

Gradient Descent: Gradient descent is an optimization algorithm used to minimize the error or loss function in machine learning models. It works by iteratively adjusting the model's parameters in the direction that reduces the error the most, based on the gradient of the loss function.

Loss Function: A loss function, also known as a cost or objective function, quantifies the difference between the predicted outputs of the model and the actual targets. The goal of training is to minimize this loss, thereby improving the model's accuracy. Common loss functions include mean squared error (for regression tasks) and cross-entropy loss (for classification tasks).

Hyperparameters: Hyperparameters are adjustable parameters that govern the training process of the machine learning model. Unlike model parameters, which are learned during training, hyperparameters are set before the learning process begins. Examples include learning rate, batch size, and the number of layers in a neural network.

Epoch: An epoch refers to one complete pass through the training dataset. During training, a model is typically exposed to the dataset multiple times, and each of these complete passes is called an epoch. The number of epochs can affect the training duration and the model's accuracy.

Bias and Variance: These are essential concepts in understanding a model's performance. *Bias* is the error introduced by approximating a complex problem with a simplistic model. *Variance* is the error introduced by the model's sensitivity to fluctuations in the training data. Achieving a balance between bias and variance is crucial for building a robust model.

Cross-Validation: This is a technique used to evaluate the performance of a machine learning model. It involves partitioning the data into subsets, training the model on some subsets, and validating it on others. This process is repeated multiple times, and the results are averaged to obtain a more reliable performance estimate.

Regularization: Regularization techniques are used to prevent overfitting by adding a penalty to the loss function based on the model's complexity. Common regularization methods include L1 (Lasso) and L2 (Ridge) regularization, which can shrink complexity-contributing parameters.

By grasping these key concepts and terminologies, you lay the groundwork for deeper exploration into the world of generative AI. Each term, while fundamental on its own, connects with others to form a cohesive framework that enables you to not only understand how machine learning models work but also to push the boundaries of what's creatively possible with these powerful tools. As we move forward, keep these foundational ideas in mind—they will serve as essential building blocks for your journey into the innovative possibilities enabled by generative AI.

CHAPTER 3:
NEURAL NETWORKS

As we delve deeper into the realm of generative AI, understanding neural networks becomes indispensable. Neural networks serve as the bedrock of modern AI, drawing inspiration from the complex web of neurons in the human brain. They're composed of layers upon layers, each designed to process data and extract features, enabling the AI to recognize patterns and make decisions. At their core, neural networks are about transforming input data into meaningful output through a series of weighted connections and activations. This transformative process allows generative AI to create stunning images, compose music, and even generate coherent text, pushing the boundaries of what's possible. By grasping the fundamental principles of neural networks, you'll be better equipped to explore their vast creative potential and drive innovations in art, technology, and beyond. The journey through this chapter will provide clarity on how these digital neurons work, paving the way for deeper exploration in subsequent chapters.

Introduction to Neural Networks

Neural networks form the backbone of modern artificial intelligence, and they're a cornerstone of machine learning. But what exactly are they? At their essence, neural networks are a series of algorithms that endeavor to recognize patterns from datasets. They interpret sensory data through a kind of machine perception, labeling, or clustering of raw input. Just as our human brain interprets data using a network of neurons, these neural networks mimic that structure but in a highly simplified and abstract manner.

The concept of neural networks isn't as new as one might think. It dates back to the 1940s, with pioneering researchers laying the groundwork for what would eventually evolve into the sophisticated models we use today. While the early neural networks were limited by the technology of their time, theoretical work continued to push forward, leading to the advanced frameworks we now employ in various applications.

A key aspect of neural networks is their structure, which consists of layers of nodes. Each node, or neuron, carries out a simple computation, and the output of these computations is passed on to the next layer. Typically, you'll find three types of layers in a neural network: input layers, hidden layers, and output layers. The input layer receives the initial data, the hidden layers perform intermediary calculations, and the output layer generates the final prediction or classification.

One might wonder how neural networks "learn". This learning process involves adjusting the weights of connections between neurons to minimize the difference between the predicted output and the actual output. This method, known as backpropagation, breaks down complex input data through multiple layers of processing, making increasingly refined predictions. It's a bit like fine-tuning a musical instrument - small adjustments can lead to better harmony.

The learning process in neural networks is often supervised, meaning the model is trained using a labeled dataset to learn the mapping between input and output. For example, a neural network trained to recognize images of cats might be shown thousands of pictures labeled as "cat" or "not cat". Over time, it adjusts its parameters to improve its accuracy in telling apart the two categories. This iterative process of learning and correction is what enables neural networks to become remarkably effective at complex tasks.

However, neural networks aren't just limited to supervised learning. They can also learn in unsupervised or semi-supervised frameworks, where they identify patterns and structures from data without explicit labels. This is particularly useful in scenarios where labeled data is scarce or expensive to obtain. For instance, clustering algorithms can group similar items together without knowing what those items are in advance, finding hidden structures within the data.

At their core, neural networks are inspired by the biological neural networks in the human brain. But while biological neurons are complex and operate electrically through the flow of ions, artificial neurons are mathematical functions that convert data inputs into outputs through weighted connections. This abstraction allows neural networks to be implemented in computers and applied to a vast range of problems, from image recognition to natural language processing.

Speaking of natural language processing, neural networks have revolutionized the way machines understand and generate human language. With the advent of more complex architectures like recurrent neural networks (RNNs) and transformers, they've enabled breakthroughs in text generation, translation, and sentiment analysis. These advanced models can capture context and semantics in ways previously unimaginable, opening up new avenues for creativity and productivity.

As we delve deeper into the world of neural networks, it becomes apparent why they're pivotal in generative AI. Generative models, such as Generative Adversarial Networks (GANs) and Variational Autoencoders (VAEs), harness the power of neural networks to create new data from learned patterns. By understanding the principles that drive neural networks, one gets a better grasp of how these generative models work and the potential they hold for creative applications.

In addition to their technical prowess, neural networks have brought about a paradigm shift in how we approach problem-solving.

They've demonstrated that with enough data and computational power, machines can achieve and sometimes surpass human-level performance in specific tasks. This has led to exciting innovations in fields as diverse as healthcare, finance, and entertainment.

Take healthcare, for example. Neural networks are being used to predict patient outcomes, classify medical images, and even assist in drug discovery. By analyzing vast amounts of data, they can uncover patterns and insights that might be missed by human experts. This not only streamlines diagnostic processes but also paves the way for personalized treatment plans, improving patient care and outcomes.

In the world of finance, neural networks are transforming the landscape through applications in fraud detection, algorithmic trading, and risk management. By learning from historical data, these models can detect anomalies, predict market trends, and make investment decisions with unprecedented accuracy. This has significant implications for enhancing financial security and optimizing investment strategies.

But the impact of neural networks isn't limited to practical applications alone. They've become a cornerstone of creative technology, enabling artists and designers to explore new forms of expression. Whether it's generating music, creating visual art, or designing interactive experiences, neural networks are pushing the boundaries of what's possible when human creativity meets machine intelligence.

As we continue to advance in this field, it's crucial to remain aware of the ethical considerations surrounding the use of neural networks. Issues of bias, transparency, and accountability must be addressed to ensure that these technologies are used responsibly and fairly. By fostering an understanding of these principles now, we can mitigate potential pitfalls and harness neural networks' transformative potential for the greater good.

In summary, neural networks are a transformative technology that has fundamentally changed our approach to problem-solving and creativity. Understanding their structure, learning mechanisms, and vast applications provides a solid foundation for anyone interested in the field of generative AI. They're not just algorithms; they're tools that mimic aspects of human intelligence, enabling machines to learn, adapt, and generate as we push the boundaries of what's possible in artificial intelligence.

Types of Neural Networks

Neural networks come in various forms and architectures, each designed to tackle specific types of problems. While all neural networks are based on the same fundamental principles, the differences in their designs and applications have a profound impact on their performance and suitability for different tasks.

One of the most straightforward types of neural networks is the **feedforward neural network**. This is the classic model most people imagine when they think of neural networks. In a feedforward network, data flows in one direction—from input to output—passing through several hidden layers in the process. These networks are commonly used for tasks like classification and regression due to their simple yet effective structure.

As the name suggests, **convolutional neural networks (CNNs)** introduce a convolutional layer that is particularly well-suited for processing grid-like data such as images. Each convolutional layer applies filters or kernels to the input, capturing features like edges, textures, and colors. This makes CNNs incredibly effective for image recognition, object detection, and even video processing tasks. Convolutional layers are followed by pooling layers to reduce the spatial dimensions, maintaining the most essential features and making the network more efficient.

On the other hand, **recurrent neural networks (RNNs)** are designed to handle sequential data, making them ideal for tasks involving time series or natural language processing. Unlike feedforward networks, RNNs have connections that form directed cycles, enabling them to maintain a 'memory' of previous inputs. This feedback loop allows RNNs to consider the temporal dynamics of data, crucial for applications like language translation, speech recognition, and text generation. However, standard RNNs sometimes struggle with long-term dependencies, where the gap between relevant information and the point of need is large.

To address this issue, variants like **Long Short-Term Memory (LSTM)** networks and **Gated Recurrent Units (GRUs)** were developed. These specialized RNNs incorporate gating mechanisms that enhance their ability to remember and forget information selectively. By managing the flow of information more effectively, LSTMs and GRUs have become popular choices for advanced sequential data tasks, from detailed text analysis to intricate time-series forecasting.

Autoencoders represent another fascinating category of neural networks aimed at unsupervised learning. These networks consist of an encoder and a decoder, which work together to compress data into a lower-dimensional representation and then reconstruct it. Autoencoders excel at tasks like data denoising, dimensionality reduction, and anomaly detection. When combined with deep learning techniques, they can even generate new data, which plays a pivotal role in various generative AI applications.

Radial Basis Function Networks (RBFNs) might not be as widely known but have unique strengths in certain applications. These networks use radial basis functions as activation functions, making them adept at classification tasks. RBFNs are typically three-layer structures with an input layer, a hidden radial basis layer, and an output layer. They excel in scenarios where interpolation in multi-dimensional space is required, offering advantages in pattern recognition and function approximation.

Pushing the boundaries further, **transformer networks** have revolutionized natural language processing tasks. Unlike RNNs, transformers do not process data in sequence; instead, they leverage an attention mechanism that allows them to weigh the significance of different parts of the input data, regardless of their position. This innovation has made models like BERT, GPT-3, and T5 highly effective at tasks like language translation, question answering, and text generation. Transformers have even made strides in image processing and other domains, showcasing their versatility.

Let's not forget **graph neural networks (GNNs)**. These networks extend the power of neural networks to graph-structured data, which is prevalent in social networks, molecule analysis, and transportation systems. GNNs can reason about the relationships and interactions between entities, making them indispensable for tasks that revolve around network analysis, recommendation systems, and even drug discovery.

Another notable category is the **generative adversarial network (GAN)**, comprised of two sub-networks—the generator and the discriminator—pitted against each other. The generator creates data samples, while the discriminator evaluates them. Through this adversarial process, GANs can generate highly realistic data, spanning applications like image synthesis, music creation, and text generation. Their ability to produce new, previously unseen data has opened up remarkable creative possibilities.

Last but certainly not least, **variational autoencoders (VAEs)** bring a probabilistic twist to the autoencoder model. VAEs encode input data into a distribution rather than a fixed point, allowing for more nuanced data representation. This probabilistic framework enables VAEs to generate new data samples similar to the training data, contributing significantly to the field of generative AI. VAEs have found applications in tasks like image and speech synthesis, providing a robust framework for creative AI endeavors.

In conclusion, the diversity of neural network types offers a rich toolkit for addressing an array of problems. From simple tasks like classification to complex generative tasks, the specialized architectures of these networks provide the means to explore and innovate in ways previously unimaginable. As you've now seen, each type of neural network brings its unique strengths to the table, opening up countless opportunities for both solving practical problems and pushing the boundaries of creativity with generative AI.

CHAPTER 4:
INTRODUCTION TO GANS

In the vibrant landscape of generative artificial intelligence, Generative Adversarial Networks (GANs) stand as one of the most groundbreaking and intriguing advancements. Conceptualized by Ian Goodfellow and his colleagues in 2014, GANs revolutionized the way machines create by pitting two neural networks against each other: a generator and a discriminator. This adversarial setup spawns a unique dynamic where the generator crafts data that could pass for real, while the discriminator aims to distinguish genuine data from the generator's fabrications. The interplay between these two networks results in strikingly realistic creations, ranging from hyper-realistic images to convincing audio deepfakes. GANs have not only pushed the boundaries of what machines can generate but have also inspired a wave of creative exploration across various fields such as art, music, and even game design. These models offer a potent glimpse into a future where algorithms collaborate as much as they compute, forging new realms of digital artistry and innovation.

What are GANs?

Generative Adversarial Networks, or GANs, are a groundbreaking framework in the realm of artificial intelligence, first introduced by Ian Goodfellow and his colleagues in 2014. Widely considered one of the most significant advancements in AI, GANs have revolutionized how we approach machine learning and data generation. Their core concept is elegantly simple yet profoundly powerful: a system of two neural networks—referred to as the generator and the discrimina-tor—locked in a game-theoretic duel to create data that is indistin-guishable from real data.

The generator's role is to produce data, whether it be images, mu-sic, or text, that mimics real-world samples. On the other side, the dis-criminator's job is to evaluate the authenticity of this data, distin-guishing between genuine and generated samples. As they compete, both networks improve, culminating in the generator becoming re-markably skilled at producing realistic data over time. This dynamic interplay is the cornerstone of GANs.

At its essence, a GAN synthesizes new data from scratch. Think about it: an AI creating a remarkably lifelike painting, composing mu-sic, or even drafting poetry that could fool human critics. This isn't just computer science wizardry; it's an evolution in creativity and in-novation. GANs empower machines to extend human imagination, producing art and artifacts previously constrained by human effort and time.

For beginners and enthusiasts, the concept might initially seem complex. To simplify, consider this analogy: imagine a forger trying to create a perfect counterfeit painting and an art expert whose sole pur-pose is to detect the counterfeit. With each new attempt from the forger, the expert becomes better at spotting fakes, and concurrently, the forger's skills improve to the point where their creations might be indistinguishable from the originals. This mutually enhancing learning process encapsulates the essence of GANs.

The potential applications of GANs stretch far and wide. In the field of computer vision, they can enhance image resolution, transform photographs into specific artistic styles, and even colorize black-and-white images. Beyond visuals, GANs extend their prowess to generating human-like speech, crafting realistic virtual environments in gaming, and revolutionizing areas like medicine by simulating sophisticated biological data for research and analysis.

But let's not overlook the fascinating element of creativity here. GANs have been at the forefront of AI-driven art, from generating surrealistic landscapes to creating portraits that challenge our perception of art itself. They offer artists and creators an entirely new medium—one where the collaboration between human intuition and machine precision results in unparalleled artistic expressions.

While the allure of GANs lies in their creative capabilities, it's also crucial to understand their underlying mechanics. The generator and discriminator both rely on neural networks, which are trained through backpropagation and gradient descent—terms you might be familiar with from earlier chapters on machine learning and neural networks. Through this training, the generator crafts data samples, attempting to deceive the discriminator. Conversely, the discriminator enhances its capability to discern real from fake until a balance is achieved.

This intricate dance is underpinned by the so-called "min-max" game. The generator aims to minimize the discriminator's ability to classify fakes correctly, while the discriminator strives to maximize its accuracy. Mathematically, this is expressed through a loss function that both networks optimize against, creating a confluence where they continually advance and improve.

One might wonder—is there a balance between achieving hyper-realism and maintaining computational efficiency? This is an ongoing area of research and innovation. For instance, researchers have developed various architectures and techniques to make GANs more efficient and versatile. Techniques such as Wasserstein GANs (WGANs) introduce modifications to the original framework to enhance training stability and improve the quality of generated outputs.

Nevertheless, the journey with GANs is filled with both excitement and challenges. One of the paramount obstacles faced by practitioners is "mode collapse," where the generator produces limited varieties of outputs, thus compromising the diversity of generated data. Various strategies, such as implementing gradient penalties and exploring different network architectures, have been employed to mitigate these issues, continually pushing the boundaries of what GANs can achieve.

What truly sets GANs apart is not just their technical prowess but their ability to democratize creativity. By providing tools that can generate art, music, and even ideas, GANs are amplifying human creativity, offering new avenues for artistic exploration and innovation. They have opened new frontiers where artists collaborate with algorithms to co-create, leading to unique and unpredictable results.

For those just starting out, the world of GANs might feel like stepping into a new dimension of possibilities. The best way to dive in is to experiment with existing tools and frameworks like TensorFlow and PyTorch, which offer pre-built GAN models. By tweaking parameters, exploring different datasets, and iterating on designs, you'll get hands-on experience with how these fascinating networks operate.

Another exciting aspect of GANs is their ability to teach us about the latent spaces within data—those abstract, multi-dimensional spaces where data is embedded and where GANs operate. Understanding these latent spaces can offer profound insights into the nature of the data itself, leading to new discoveries and creative breakthroughs.

The beauty of GANs lies in their dual nature—the continual tug-of-war that leads to ever-improving results, and the boundless creative potential they unlock. Whether it's in generating photorealistic images or crafting symphonies that resonate with human emotion, GANs are a testament to the incredible strides artificial intelligence has made.

So, as you continue this journey into generative AI, let the concept of GANs spark your curiosity and ignite your imagination. Explore, iterate, and innovate—your next masterpiece could just be a dataset and a few lines of code away.

In future sections, we'll dive deeper into how GANs work, and you'll get a chance to see practical applications and advanced techniques. But for now, take a moment to appreciate the elegance and power of GANs. They're not just a tool—they're a gateway into new realms of creativity and expression.

How GANs Work

To understand how GANs, or Generative Adversarial Networks, work, we first need to unpack the unique architecture that underpins them. Think of a GAN as a sort of digital tug-of-war between two different neural networks: the generator and the discriminator. Both of these components play distinct yet complementary roles, pushing each other towards improvement with every iteration. This interactive dynamic creates an environment conducive to generating impressively realistic data.

The generator's job is to create synthetic data that mimics real-world data as closely as possible. Initially, the generator starts off with producing random noise, which is essentially meaningless. As the training progresses, the generator learns to produce data that is increasingly similar to the real data it's modeled after. The discriminator, on the other hand, is like the quality controller. It evaluates the data coming from the generator against the real data, striving to distinguish between the two.

This interaction between the generator and discriminator is akin to a forger and an art critic. The generator, or forger, tries to create art pieces that look as genuine as possible, while the discriminator, or art critic, evaluates each piece to see if it's authentic or a forgery. Over time, both improve—the forger becomes more skilled at creating convincing pieces, and the critic becomes better at spotting fakes.

The training process can be broken down into a few key steps:

- **Initialize the Networks:** Both the generator and discriminator are initialized with random weights.

- **Generate Data:** The generator creates a batch of data based on random noise or some other initial input.

- **Discriminate Data:** The discriminator evaluates two batches of data—one real and one generated. It tries to determine which is which.

- **Calculate Loss:** Both networks calculate their respective losses. For the discriminator, it's a measure of how well it can distinguish between real and fake data. For the generator, it's how well it can 'fool' the discriminator.

- **Update Weights:** Using backpropagation, both networks update their weights to minimize their losses. The generator adjusts its weights to produce more convincing data, while the discriminator fine-tunes its ability to detect fakes.

- **Iterate:** These steps are repeated for many iterations or epochs, gradually improving the generator's output and the discriminator's accuracy.

But here's where it gets particularly fascinating: this adversarial relationship creates a feedback loop that drives continuous improvement. Initially, the generated data may be easily identified as fake, but as the generator learns, it produces increasingly convincing data. Likewise, the discriminator becomes more adept at spotting subtle flaws.

A pivotal aspect of this process is the concept of the loss function, specifically the minimax game. The discriminator aims to maximize its accuracy in distinguishing real from fake data, while the generator aims

Chris Elliott

to minimize the discriminator's ability to do so. This dynamic creates a balance—both networks improve together, pushing each other to enhance their performance.

The generator's loss function focuses on how poorly it performs in tricking the discriminator. Conversely, the discriminator's loss function centers on its accuracy in differentiating between real and fake data. The objective for the generator is to produce data so realistic that the discriminator's accuracy drops to the level of random guessing.

In practice, training GANs can be quite challenging. One common issue is maintaining the balance between the generator and the discriminator. If one network becomes too strong, it can overpower the other, leading to suboptimal training. For example, if the discriminator becomes too accurate too quickly, the generator may struggle to improve, as it receives overly harsh feedback. Similarly, if the generator improves too fast, the discriminator may never catch up, reducing its effectiveness.

To address this, researchers often experiment with various techniques, such as modifying the architecture or altering the training algorithms. For instance, some approaches involve training the generator more frequently than the discriminator, or vice versa, to maintain equilibrium. Others might tweak the loss functions to provide smoother gradients, facilitating more effective learning.

Beyond these fundamentals, there are also numerous variations of GANs tailored for specific applications. For example, Conditional GANs (cGANs) incorporate additional information, like class labels, into the training process. This allows the generator to produce data that adheres to certain conditions or classifications, adding a layer of control over the output.

Another notable variant is the Deep Convolutional GAN (DCGAN), which utilizes deep convolutional neural networks in both the generator and discriminator. This architecture is particularly effective at handling image data, resulting in higher-quality image generation. Similarly, Progressive Growing GANs (PGGANs) gradually increase the complexity of the generated data over the training process, often yielding better results for high-resolution images.

The possibilities with GANs extend far beyond images. They've been employed in numerous creative and practical applications, from music composition to game design. For instance, GANs can generate unique musical compositions by learning from vast collections of existing music. In game development, GANs help in creating more dynamic and realistic environments, enhancing the overall gaming experience.

Despite their powerful capabilities, GANs are not without their challenges. Overfitting, mode collapse, and convergence difficulty are some common hurdles researchers encounter. Overfitting occurs when the model performs exceptionally well on training data but poorly on unseen data. Mode collapse happens when the generator produces limited variations in data, reducing diversity. Convergence difficulty refers to the challenge of achieving stable and efficient training, as the adversarial nature of GANs can sometimes lead to oscillations or divergent behavior.

Addressing these challenges requires a blend of theoretical understanding and practical experimentation. Researchers continually explore new techniques to stabilize training, enhance generalization, and increase diversity in generated data. Techniques like spectral normalization, Wasserstein GANs, and data augmentation are some methods that show promise in overcoming these obstacles.

As GANs continue to evolve, their potential applications are expanding. They're playing a crucial role in advancing fields like computer vision, natural language processing, and even medical research. For example, GANs are being used to enhance medical imaging, aiding in the diagnosis and treatment of diseases by generating high-resolution images from low-resolution scans.

Ultimately, the significant strides in GAN technology underscore the importance of understanding how these systems work. By delving into the intricacies of the generator-discriminator dynamic, we can better harness GANs' potential, driving innovation across numerous domains. This foundational knowledge also empowers individuals to explore their creative potential, using GANs to produce unique works of art, music, and more.

In conclusion, grasping the mechanics of GANs opens up a world of possibilities. From generating stunning visuals to composing new music, the applications are as diverse as they are exciting. As we continue to refine these systems, the boundary between the real and the synthetic will blur, heralding a new era of creativity fueled by artificial intelligence

CHAPTER 5:
INTRODUCTION TO VAES

Venturing into the heart of generative artificial intelligence, we find Variational Autoencoders (VAEs), a technique that melds the richness of deep learning with probabilistic inference. VAEs offer a fascinating approach to generating data that is remarkably realistic by learning intricate structures of the training data, capturing nuances that other models might miss. Unlike the adversarial nature of GANs, VAEs build on the foundations of autoencoders but introduce a probabilistic twist, which enables them to handle the uncertain nature of real-world data more effectively. This chapter will unravel how VAEs use latent variables to create more flexible and robust generative models, opening doors to inventive and striking applications in fields like art, music, and beyond. Here, you'll discover the key principles behind VAEs and why they are pivotal in pushing the boundaries of creativity and innovation in AI.

What are VAEs?

Variational Autoencoders, or VAEs, are a class of generative models in machine learning with a compelling blend of theoretical elegance and practical utility. At their core, VAEs aim to encode data in a continuous and compact form, which can then be used to generate new, similar data. Imagine taking a complex, high-dimensional dataset—like images of handwritten digits—and distilling it into a set of simpler, meaningful variables. This distilled representation, or "latent space," is where the magic happens with VAEs.

Chris Elliott

Unlike traditional autoencoders, which compress data into a fixed representation, VAEs introduce a probabilistic twist. This probabilistic nature is fundamental because it allows for the generation of new data points. Essentially, a VAE doesn't just encode each input into a single point in the latent space. Instead, it maps the input to a distribution—usually Gaussian—over the latent space. This means every input is encoded not just as a point but as a cloud of possibilities.

Why is this important? Well, when you generate new data, you're actually sampling from this distribution. This approach leads to a more robust and diverse set of generated outcomes, which is particularly valuable in creative applications. You're not merely copying or slightly tweaking existing data; you're crafting new instances with novel qualities. The continuous nature of the latent space in VAEs also offers smooth interpolation between data points, which can be aesthetically pleasing and richly informative.

So, what makes VAEs so special? It's their dual role of being both an encoder and a decoder, operating with a probabilistic mindset. The encoder takes in data and learns the underlying distributions in the latent space. This helps the model understand the essence of what it's encoding. It's a bit like taking an intricate piece of art and understanding the brush strokes, the hues, the subtleties—elements that make it unique. The decoder then samples from this latent space to generate new data, akin to an artist using those learned techniques to create new works.

To make this concrete, let's consider the example of generating handwritten digits, a popular test case for VAEs. The VAE learns to encode various digit images into a latent space where similar digits are clustered close together. By exploring points in this latent space, you can generate new digit images that may not exist in the training data but still appear authentic. This generative capability is what sets VAEs apart as a tool for both data compression and creative exploration.

But VAEs aren't just limited to images. Their applications span a wide range of domains, from text generation to music composition. In essence, any form of data that can be meaningfully compressed and then expanded can benefit from the VAE approach. The probabilistic foundation also makes VAEs robust in handling noisy and incomplete data, which is often the case in real-world scenarios. For instance, in medical imaging, VAEs can reconstruct high-quality images from partial scans, which can be crucial for diagnostics.

Technically, the probabilistic framework of VAEs stems from variational inference, a method in Bayesian statistics. In layman's terms, variational inference approximates complex probability distributions with simpler ones, making it computationally feasible to perform tasks like encoding and decoding. The objective function, known as the evidence lower bound (ELBO), balances two key goals: the accuracy of data reconstruction and the alignment of the latent distribution with a prior distribution (usually Gaussian). This balance ensures that the latent space is both informative and structured.

From a mathematical perspective, the process can be broken down into a few crucial steps. First, the encoder network, also known as the recognition model, projects the input data into the latent space, producing both a mean and a standard deviation for the underlying Gaussian distribution. The next step involves sampling from this distribution to produce a latent variable. This step, known as the "reparameterization trick," is essential to make the process differentiable, allowing the model to be trained using gradient descent. Finally, the decoder network, or the generative model, reconstructs the input data from the latent variable.

Despite its mathematical rigor, the practical implementations of VAEs are surprisingly accessible. Popular machine learning frameworks like TensorFlow and PyTorch offer pre-built components for constructing and training VAEs, making it easier even for beginners to experiment with these models. The ability to tweak parameters and explore various architectures adds an element of creativity to the technical challenge, making it not just an exercise in coding but also in design.

One of the enriching aspects of learning about VAEs is understanding the interpretability of the latent space. This space often reveals insightful patterns and structures inherent in the data. For instance, in facial image generation, different dimensions of the latent space might correspond to features like hair color, angle of the face, or even emotional expression. Exploring these dimensions offers a rich, hands-on experience of interacting with complex data in a more intuitive, visual way.

Because VAEs are generative models, they're naturally suited for creative fields. Artists and designers can use VAEs to explore new aesthetic possibilities, generating novel artworks, design patterns, or even fashion items. The fusion of creativity and technicality in VAEs opens up a playground where art and science meet. Furthermore, the smooth interpolation in the latent space allows for the creation of artistic transitions and morphs, providing a dynamic and engaging medium for storytelling and artistic expression.

Moreover, the concepts behind VAEs can be extended and modified for specialized applications. Conditional VAEs (CVAEs), for instance, incorporate additional information along with the input data, allowing for more controlled and context-aware generation. This has major implications in fields like text-to-image synthesis, where the generated image needs to correspond to specific textual descriptions. The flexibility and extensibility of VAEs make them a powerful tool in the generative AI toolkit.

The journey into understanding VAEs is as much about theory as it is about application. It's about connecting abstract mathematical ideas with concrete, tangible outputs. As you delve deeper into this subject, you'll find that VAEs offer a unique vantage point into the generative process, blending rigor with creativity. The potential applications are vast, limited only by imagination and innovation.

In summary, VAEs serve as a fascinating intersection of data compression and generative modeling, offering a potent mix of probabilistic embedding and reconstruction. Their utility spans multiple domains, making them versatile tools for both scientific and artistic endeavors. By harnessing the strengths of variational inference and neural networks, VAEs create a rich, continuous latent space that fosters creativity and exploration. Whether you're looking to generate lifelike images, compose compelling music, or understand complex datasets, VAEs stand out as a transformative technology in the realm of generative AI.

How VAEs Work

To grasp how Variational Autoencoders (VAEs) work, it's key to first understand that they serve two primary purposes: compression and generation. Unlike standard autoencoders, which focus mainly on compressing data, VAEs add a twist by incorporating probabilistic elements. This allows them to not only recreate input data but also generate new, meaningful instances. The dual capabilities of VAEs make them particularly useful in creative fields like digital art and music where the generation of new content is invaluable.

The architecture of a VAE consists of two main components: the encoder and the decoder. The encoder takes input data, such as an image or text, and transforms it into a compact, often lower-dimensional representation called the latent space. The trick here is that this transformation is not deterministic. Instead, it encodes inputs as distributions over the latent space, represented typically using Gaussian distributions. This probabilistic nature is what allows VAEs to generate new data that is both varied and coherent.

Now, let's dive deeper into the structure of the latent space. The latent space in a VAE is a continuous, multidimensional vector space. Each point in this space corresponds to a potential output of the network. During the encoding process, the encoder outputs two vectors: a mean vector and a standard deviation vector. Together, these define a Gaussian distribution in the latent space for each input sample. The process involves sampling a latent vector from this distribution, which is then fed into the decoder to reconstruct the original data.

Employing the reparameterization trick is crucial for training VAEs. This technique allows for backpropagation, essential for optimizing neural networks. Essentially, instead of sampling directly from the distribution defined by the encoder, the model samples a vector from a standard Gaussian distribution and then scales and shifts it according to the mean and standard deviation parameters output by the encoder. This clever maneuver enables the gradient calculations needed to efficiently train the model.

The decoder's job is to transform the sampled latent vector back into the original data. For image data, this means turning a compact latent vector back into a full-resolution image. Stochastic variations in the latent space translate to diverse, yet plausible, outputs. This capability opens up a world of creative possibilities, from generating new art pieces to designing unique virtual objects.

It is vital to understand the role of the loss function in VAEs. The loss function has two main components: the reconstruction loss and the KL divergence. The reconstruction loss measures how well the decoder manages to recreate the input data. Typically, it's computed using metrics like mean squared error for numerical data or binary cross-entropy for image data. The KL divergence term, on the other hand, enforces the distributions in the latent space to follow a standard normal distribution. This regularization ensures that the latent space is well-behaved and lets us generate meaningful and coherent data samples.

The unique loss function of a VAE serves an essential role in balancing the trade-off between reconstruction quality and the regularity of the latent space. Without this balance, the VAE might overfit to the training data, losing its ability to generate new, varied outputs. Such a predicament would limit the VAE's utility in creative applications where generating diverse content is crucial.

Furthermore, VAEs can easily be extended to conditional VAEs (CVAEs), where the model is conditioned on additional information or labels. In a CVAE, both the encoder and decoder also take class labels or other forms of conditional data as input. This conditioning enables more controlled generation, such as creating art that adheres to specific styles or genres. The flexibility introduced by conditional elements enriches the scope of creative applications even further, offering artists and developers detailed control over the outputs.

Training VAEs is computationally intensive, requiring potent hardware and substantial data. Nevertheless, the benefits outweigh the challenges when these models produce astonishingly novel and high-quality outputs. Many modern graphics processing units (GPUs) come equipped with the necessary computational power to train VAEs efficiently.

Moreover, the intersection of VAEs with other generative models such as GANs provides even more intriguing possibilities. Some advanced models combine the strengths of both VAEs and GANs, leveraging the stability and interpretability of VAEs while capitalizing on the sharp generation capabilities of GANs. These hybrid models push the boundaries of what's possible in generative tasks.

It's also worth mentioning how user-friendly certain programming libraries have made the task of implementing VAEs. Popular frameworks like TensorFlow and PyTorch offer pre-built functions and modules specifically for VAEs, simplifying the development process. This ease of use democratizes access to this powerful technology, allowing beginners and enthusiasts to dive into generative AI without requiring an extensive background in machine learning.

Looking ahead, the limitations and challenges faced by VAEs today are being actively addressed by the research community. Whether it's improving the fidelity of generated images, reducing training times, or enhancing the stability of the models, continuous advancements are making VAEs an even more attractive option for creative and practical applications alike.

Unquestionably, understanding the inner workings of VAEs provides not only a glimpse into their potential but also a solid foundation for exploring the broader landscape of generative AI. As you delve deeper into this world, the principles behind VAEs will serve as a cornerstone, enabling you to build more complex and exciting projects.

Chapter 6:
Other Generative Models

Diving beyond GANs and VAEs, we explore a diverse spectrum of generative models that offer unique ways to create and innovate. Autoregressive models, like PixelRNN, build data point by data point, delivering high-quality sequences by predicting future values based on prior ones. Meanwhile, flow-based models employ reversible transformations to map data to latent spaces, providing exact density modeling and efficient sampling through architectures like RealNVP. These models not only expand the toolbox for artists and developers but also open new avenues for experimentation and discovery, transcending the boundaries of traditional generative methods and pushing the frontiers of AI creativity.

Autoregressive Models

Autoregressive models are a fascinating subset of generative models that have piqued the interest of AI enthusiasts and researchers alike. These models predict the next element in a sequence based on the previous elements, which makes them incredibly useful in a variety of applications ranging from text generation to music composition. At their core, autoregressive models harness the power of probability distributions to make informed predictions.

The foundational idea behind autoregressive models is relatively straightforward. Imagine you're trying to predict the next word in a sentence. The model will examine the words that precede the target word and generate probabilities for possible next words based on learned patterns. Essentially, it's engaging in a probabilistic guessing game, but one backed by intricate learning and vast data.

One of the most well-known applications of autoregressive models is in natural language processing (NLP). OpenAI's GPT (Generative Pre-trained Transformer) models are prime examples. These models have demonstrated the ability to create coherent, contextually relevant text, making them invaluable for applications like chatbots, writing assistants, and automated content creation. Their ability to perform tasks such as summarization, translation, and even creative writing showcases their versatility.

Be it text, music, or even images, autoregressive models are not limited to a single domain. For instance, in the realm of music generation, these models can predict the next note in a melody or even an entire sequence of chords. By training on vast corpora of musical compositions, they learn the stylistic and structural nuances of different genres, enabling them to produce original pieces that echo the characteristics of existing genres or create entirely new ones. This opens up endless possibilities for musicians and composers to explore new creative dimensions.

Diving deeper into the mechanics, autoregressive models are trained using a method called "maximum likelihood estimation". They learn to maximize the probability of the training data by adjusting their internal parameters. This learning process involves backpropagation and optimization techniques that iteratively refine the model's weights, allowing it to better predict future elements in a sequence.

One of the strengths of autoregressive models is their simplicity and effectiveness. Unlike some other generative models that require complex architectures and training processes, autoregressive models are relatively straightforward to implement and train. This makes them accessible to beginners and experts alike, providing a robust foundation for exploring generative AI.

However, it's important to acknowledge the challenges associated with autoregressive models. One notable issue is their tendency to accumulate errors over long sequences. Because each prediction depends on the previous ones, any mistakes can propagate and compound, leading to less coherent outputs as the sequence lengthens. Researchers are continuously working on improving techniques to mitigate these errors and enhance the overall quality of generated sequences.

Another potential pitfall is the immense computational resources required for training large-scale autoregressive models, especially those used in NLP. The sheer volume of data and the complexity of calculations necessitate significant processing power and memory. While this might be a constraint for some, advancements in hardware and more efficient algorithms are gradually alleviating these barriers, making high-performance models more accessible to a broader range of users.

We must also consider the creative potential that autoregressive models unlock. For artists and creators, these models can serve as collaborative tools, offering new perspectives and ideas that may have been otherwise unexplored. Imagine a writer using an autoregressive model to draft the next chapter of a novel, using the model's suggestions to overcome writer's block and experiment with different plot twists. Or a musician leveraging generated harmonic sequences to inspire new compositions, pushing the boundaries of traditional music theory.

In conclusion, autoregressive models represent a cornerstone of generative AI, offering powerful capabilities for sequence prediction across diverse domains. By leveraging the power of probability and learning from vast datasets, these models are able to generate convincing, contextually relevant sequences that extend human creativity. The journey doesn't end here, though. With continuous advancements in

AI research and technology, the potential for autoregressive models to transform the creative landscape is boundless and truly inspiring.

Flow-Based Models

Flow-based models represent a unique approach within the realm of generative models. These models primarily revolve around the concept of invertible neural networks, where the mapping between input data and latent variables is bidirectional. This bidirectionality ensures that the model maintains a perfect bijective mapping, meaning each input data point has a unique corresponding point in latent space, and vice versa. It's this characteristic that sets flow-based models apart from other generative models such as GANs and VAEs.

One of the foundational principles of flow-based models is the sequence of invertible transformations. These transformations are applied to input data to map it onto a simpler distribution, commonly a Gaussian distribution. Because these transformations are invertible, they allow for both sampling new data from the model and computing the exact probability density function of the data, something not readily possible with many other generative models. This capability opens up new avenues for tasks that require precise density estimation.

The backbone of flow-based models can be described as a chain of simple, invertible transformations that progressively warp the data distribution into the desired form. Each of these transformations typically involves a combination of scaling, rotation, and translation operations, making sure that the entire process remains invertible. The most well-known architectures within this family of models include Real-NVP (Real-valued Non-Volume Preserving transformations) and its successors, such as Glow.

RealNVP laid the groundwork for understanding how to design these invertible transformations effectively. By splitting the input data into different parts and applying separate transformations to each, it manages to control the complexity while ensuring that inversion remains computationally feasible. This strategy allows RealNVP to achieve impressive results on image generation tasks, providing high-quality, realistic samples.

Glow, another notable flow-based model, extends the ideas presented in RealNVP and adds significant optimizations. For example, Glow introduces a new type of flow layer called the "1x1 invertible convolution." This layer improves the expressiveness and flexibility of the model, allowing it to achieve even better performance on tasks such as image synthesis. Additionally, Glow implements a more efficient parameterization and training algorithm, further enhancing its scalability and applicability to larger datasets.

The ease of computing the likelihood directly in flow-based models is a substantial advantage. Because these models allow for exact density computation, they can be particularly useful in applications where understanding or manipulating the underlying data distribution is essential. For instance, in anomaly detection, knowing the precise likelihood of each data point helps identify outliers more effectively. Similarly, in scenarios requiring reliable uncertainty quantification, flow-based models can provide exact probabilistic estimates.

Another advantage of flow-based models is their robustness in handling high-dimensional data. These models are inherently structured to deal with complex data distributions, making them suitable for applications in image and audio synthesis. By leveraging invertible transformations, flow-based models can capture intricate details and dependencies within the data, producing high-fidelity outputs that enhance the creative potential of generative AI.

Despite their strengths, flow-based models also come with certain challenges. The computational cost of training such models can be high, primarily because the inverse transformations require careful design and efficient implementation. This makes training slower compared to some other generative models, particularly on large-scale datasets. Additionally, the architecture of flow-based models can be complex, requiring a deep understanding of invertible functions and numerical stability issues during implementation.

A critical consideration when working with flow-based models is the choice of transformations used within the network. These transformations need to strike a balance between being expressive enough to capture complex data distributions and being tractable for inversion. Researchers continue to explore new types of invertible transformations and optimization techniques to address these challenges.

Furthermore, the design space for flow-based models is quite vast and still under active exploration. There is ongoing research into how to best structure these models and the types of flows that offer the best trade-offs between complexity and performance. Innovations in this area could make flow-based models even more powerful and accessible, lowering the barriers for enthusiasts and beginners interested in generative AI.

In summary, flow-based models hold a unique position in the constellation of generative models thanks to their invertibility and direct likelihood computation. They offer precise control over data generation and enable tasks that require exact density estimates. While they come with their own set of challenges, the evolving research and continued improvements in their architecture make them a fascinating and valuable tool for anyone delving into the world of generative AI. Exploring flow-based models can provide deep insights into how data can be transformed and represented, offering new perspectives on both practical applications and creative possibilities.

CHAPTER 7:
TOOLS AND SOFTWARE FOR
GENERATIVE AI

In the realm of generative AI, the right tools and software can be the keys to unlocking your creative potential. From user-friendly libraries like TensorFlow and PyTorch to specialized platforms such as RunwayML, there's a rich ecosystem designed to support both beginners and seasoned developers. These tools not only simplify complex processes like neural network training but also offer pre-trained models that can be fine-tuned to your specific needs. Setting up your workspace often involves configuring environments in Jupyter Notebooks or leveraging cloud-based resources like Google Colab, making high-performance computation accessible without heavy investments. By acquainting yourself with these essential tools, you're not just learning to use software—you're embarking on a journey to translate abstract algorithms into tangible, awe-inspiring creations.

Popular Tools and Libraries

When diving into the fascinating world of generative AI, having the right set of tools and libraries can make all the difference. Whether you're creating visual art, generating music, or experimenting with text, specific software and frameworks have been designed to streamline the development process and unleash your creative potential. Each tool offers a unique set of features, making it easier to implement complex algorithms without diving too deep into the technical mumbo-jumbo.

Let's start with **TensorFlow**. Developed by Google Brain, this open-source library is immensely popular among AI researchers and developers. TensorFlow offers an extensive suite of tools for building and training machine learning models. Its versatility and comprehensive documentation make it a favorite for developing everything from deep neural networks to advanced generative models. Furthermore, TensorFlow provides TensorFlow.js, which allows you to bring ML models into the realm of JavaScript, opening up a whole new world of possibilities for web-based applications.

PyTorch is another heavyweight in the realm of generative AI. Developed by Facebook's AI Research lab (FAIR), PyTorch has gained traction for its ease of use and flexible architecture. Unlike TensorFlow, which was initially critiqued for being more complex to learn, PyTorch allows for dynamic computation graphs, making it easier to experiment and debug models. Its Pythonic nature also means that it's incredibly intuitive for Python developers. PyTorch's integration with powerful libraries like torchvision and torchaudio makes it a compelling choice for generating visual art and music.

The **Keras** library, which is now part of the TensorFlow ecosystem, is designed to be user-friendly and versatile. It offers a high-level API that makes it simpler to develop neural networks, abstracting much of the boilerplate code that often bogs down machine learning projects. Keras is particularly popular for quick prototyping and is widely used in academia and industry for both research and production purposes. One of the key advantages of Keras is its modularity – you can plug and play various components, such as loss functions, optimizers, and layers, to design custom models with minimal fuss.

GANs (Generative Adversarial Networks) have also seen significant contributions in terms of specialized libraries. One such library is

GANLab, which provides an interactive, visual tool specifically designed to help users understand the intricacies of GANs. By offering a hands-on approach, GANLab allows you to experiment with different parameters and visually observe how GAN models learn and generate data. This can be particularly beneficial for beginners who want to get a more intuitive grasp of the often perplexing world of adversarial networks.

Shifting gears to **natural language processing (NLP)**, libraries like **Transformers** by Hugging Face have become indispensable. Transformers provide state-of-the-art pre-trained models for NLP tasks such as text generation, translation, and summarization. The library abstracts much of the complexity involved in setting up and fine-tuning large language models, allowing developers to focus more on creative applications rather than the intricacies of model architecture.

For those focused on creating visual art, **RunwayML** stands out as a particularly user-friendly platform. RunwayML allows artists and designers to use AI without writing any code, providing a graphical interface to apply various machine learning models. It supports a variety of pretrained models that can be used for tasks ranging from image synthesis to style transfer. The seamless integration with creative software like Adobe Photoshop further enhances its appeal, making it easier than ever to blend AI-generated elements with traditional design practices.

Magenta is another noteworthy project, developed by Google's Brain team, that focuses specifically on the intersection of machine learning and the arts. Designed to facilitate the creation of music, video, images, and other forms of art, Magenta offers various open-source models, tools, and datasets. Its library Magenta.js even extends these capabilities to the web, making it possible to incorporate procedurally generated content into websites and interactive installations.

The development of AI art also benefits from platforms like **Artbreeder**. Built upon GANs, Artbreeder allows users to create images by mixing different pictures and adjusting various parameters such as style and color. While it might appear simplistic on the surface, Artbreeder offers a highly interactive experience for generating art, giving users an intuitive way to explore different artistic possibilities.

Another essential resource is **OpenAI**, known for developing cutting-edge models like GPT-3. With the power to generate human-like text, OpenAI's models have vast implications for creative writing, game design, chatbots, and more. Although the API for GPT-3 is not open-source, developers can leverage it through a variety of integration options, making it easier to experiment with advanced NLP functionalities without needing large-scale computational resources.

In the realm of music generation, tools like **Ableton's Live** software provide integration options with AI-based plugins. These plugins can assist in generating melodies, harmonies, and even entire compositions. Similarly, **Spleeter** by Deezer is an open-source tool that uses deep learning to separate stems from music tracks, allowing for creative remixes and mashups.

Beyond the well-known libraries and tools, there are also niche frameworks designed for specific creative applications. For example, **Processing** and **p5.js** offer visually oriented coding environments that are particularly popular in the realm of interactive art and design. These tools provide a canvas for programmers and artists alike to develop algorithmic art, visualize data, or experiment with generative graphics.

Finally, it's worth mentioning platforms like Kaggle, which provide access to datasets and a community of like-minded individuals passionate about AI and machine learning. Kaggle also offers kernels, notebooks, and competitions that can serve as practical tutorials, helping you to learn new skills and apply your knowledge in meaningful projects.

In summary, the vast array of tools and libraries available can significantly lower the barrier to entry in the field of generative AI. By leveraging these resources, beginners and enthusiasts can focus on the creative possibilities rather than getting bogged down by the technical details. The combination of user-friendly interfaces, comprehensive documentation, and robust communities makes it easier than ever to embark on a journey into the exciting world of generative AI.

Setting Up Your Workspace

Before diving into the creative and technical processes of generative AI, it's essential to establish a well-organized workspace. An efficient workspace doesn't just facilitate smoother workflow; it also enhances creativity and experimentation. With the right tools, you can focus more on creating and less on troubleshooting technical issues.

Your workspace setup will vary depending on your objectives. Are you focused on deep learning models, dabbling in neural networks, or interested in generating art and music? Each area has its own set of requirements. However, some fundamental elements apply universally, which we'll cover here.

First and foremost, let's talk about hardware. Your computer's specifications will play a crucial role in determining how efficiently you can run complex models. Basic requirements include a modern, multi-core processor, at least 16GB of RAM, and a high-end GPU. NVIDIA GPUs are often preferred for machine learning tasks because they support CUDA, a parallel computing platform that significantly speeds up training times.

Next up is your operating system. Both Windows and macOS are capable of handling generative AI tasks, but many find that Linux offers greater flexibility and stability for heavy computational tasks. Ubuntu, a popular Linux distribution, is widely recommended for machine learning enthusiasts. That said, don't feel restricted; use which-ever operating system you're most comfortable with.

After ensuring your hardware is up to par, the next step is to install some essential software. A robust development environment like Anaconda simplifies package management and deployment. Anaconda includes Python, Jupyter Notebooks, and various other pre-installed libraries crucial for generative AI. Jupyter Notebooks, in particular, offer an interactive way to write and debug code, which can be incredibly beneficial when experimenting with different models and algorithms.

Python remains the dominant language for generative AI and ma-chine learning projects. Thus, installing Python should be one of your first actions. Depending on your interests, you might also consider other languages like R or Julia, but Python's extensive library support and community make it difficult to beat.

Speaking of libraries, some essential Python libraries you'll likely need include TensorFlow, Keras, and PyTorch for building and train-ing neural networks. These libraries help abstract some of the com-plexities associated with neural networks, making it easier for beginners to get started. Additionally, they come with extensive documentation and community support, which can be invaluable.

You'll also want to install libraries specifically for data manipula-tion and visualization like NumPy, Pandas, and Matplotlib. These tools will assist you in cleaning and preparing your dataset, as well as visualizing the results, thereby providing a more intuitive understand-ing of your model's performance.

For those inclined towards visual projects, software like OpenCV can be beneficial for image and video processing. Similarly, for text and Natural Language Processing (NLP) tasks, libraries like NLTK or SpaCy will be indispensable. These tools provide pre-trained models and functions to simplify tasks such as tokenization, entity recognition, and sentiment analysis.

Don't forget about version control systems. Git is crucial for tracking changes, collaborating with others, and managing different versions of your projects. GitHub or GitLab can be used to host your repositories, and both offer free tier options suitable for individual projects. Additionally, learning basic Git commands will provide you with a valuable skill set useful in any coding discipline.

Finally, consider utilizing Integrated Development Environments (IDEs) like PyCharm or Visual Studio Code. These tools offer code completion, debugging support, and extensions that can drastically improve your coding efficiency and reduce errors. Many IDEs also integrate seamlessly with version control systems, so you can push updates to your Git repository without leaving your coding environment.

Setting up a virtual environment is another good practice. Virtual environments help you manage dependencies for different projects without causing conflicts. You'll find this particularly useful when juggling multiple generative AI projects, each requiring different packages or library versions.

Cloud computing is another avenue worth exploring, especially if you lack the local hardware to run intensive models. Platforms like AWS, Google Cloud, and Microsoft Azure offer machine learning services with powerful GPUs, and many also provide free tiers for limited usage. Utilizing cloud services can be cost-effective and offers scalability that local hardware setups cannot match.

If you're working in a team or planning to share your projects, collaboration tools like Slack, Microsoft Teams, or GitHub Projects can be beneficial. These platforms allow for real-time communication, task management, and improved coordination among team members, making collaborative efforts seamless.

Remember, setting up your workspace effectively is not a one-size-fits-all process. It depends significantly on your specific focus and project requirements. While the guidelines mentioned here are starting points, don't hesitate to customize your setup as you go along. Embracing flexibility will allow you to adapt to new challenges and opportunities, thereby optimizing both productivity and creativity in your generative AI endeavors.

There's a certain artistry to curating your workspace. It's the foundation upon which your future AI masterpieces will be built. Invest the time now to create a setup that will serve you well across the journey, and you'll thank yourself later. Whether it's tweaking your GPU settings or learning that one extra Git command, every little bit contributes to a smoother, more enjoyable experience in the world of generative AI.

The right workspace won't make you a genius overnight, but it will facilitate the rigorous and exciting process of becoming proficient in generative AI. So go ahead, assess your needs, get your hardware and software in order, and set the stage for endless creative possibilities. Once your workspace is ready, you'll be well on your way to exploring, experimenting, and crafting incredible generative art pieces.

CHAPTER 8:
DATA FOR AI ART

To truly harness the power of AI in art, understanding the role of data is paramount. Data is the backbone that drives the creativity of generative models, turning raw information into mesmerizing visuals. Collecting and preparing this data involves sourcing diverse datasets that can fuel the AI's understanding of patterns, textures, and compositions. Whether it's curating thousands of images, tweaking metadata, or balancing datasets to avoid biases, each step shapes the outcome of the AI's artistic endeavors. However, it's crucial to navigate these waters with a keen awareness of ethical considerations, ensuring that data sources are used responsibly and creators' rights are respected. By mastering the intricacies of data collection and preparation, you set the stage for AI to push the boundaries of artistic expression, crafting works that are as innovative as they are beautiful.

Data Collection and Preparation

In the realm of AI art, the quality of the data you collect and how you prepare it are crucial steps that set the foundation for your generative models. It's a bit like gathering paints and brushes before you start a painting. Without the right materials, your creation might fall flat. Let's delve into the meticulous process of data collection and preparation, ensuring that your AI art projects have the best possible starting point.

Before diving into the technical details, it's essential to understand what kind of data is useful for AI art. Commonly, the data involves images, but it might also include sound, text, or even a mix of these. The first step is to define the scope of your project and identify the specific type of data that will be most effective. For instance, if you're looking at generating abstract art, you might opt for a dataset rich in diverse visual styles. If your focus is on creating music, you'd look for audio samples.

Once you've identified the type of data you need, the next step is to source it. This is easier said than done. You'll need to scout various repositories, online databases, and even consider creating your own dataset. Some popular sources for image data include websites like Unsplash, Google Images, and specialized databases like the ImageNet. For text, repositories like Project Gutenberg or language datasets from Kaggle could be very useful. Sound can be sourced from free sound libraries or pre-existing collections of music snippets.

The ethical considerations of data collection can't be overstated. Respecting copyright and privacy laws is paramount. In many cases, it's advisable to use publicly available or open-source datasets. When you can't avoid using copyrighted material, ensure you have the necessary permissions and clearly understand the licensing agreements associated with it. Also, consider the ethical implications of the data you're using. Be mindful of biases and ensure your dataset represents a fair and accurate distribution.

After gathering your data, the preparation phase begins. This stage is less glamorous but equally crucial. Raw data is almost never in the perfect format for training models. You need to clean, filter, and potentially augment it to make it useful. Cleaning the data might involve removing duplicates, correcting errors, or discarding irrelevant samples. Filtering allows you to focus on the most representative samples, increasing the quality of your training set.

For image datasets, data augmentation techniques can be especially helpful. These methods include random cropping, flipping, rotation, and color adjustments. The goal here is to make your dataset more robust by introducing variations that the model might encounter in real-world scenarios. This way, your model won't just learn to reproduce the exact images but will grasp the underlying patterns and structures.

Normalization is another critical step in data preparation. Different datasets might have varying ranges of values, which can confuse the model during training. Normalizing the data ensures that each feature contributes equally to the learning process. For images, this often involves changing pixel values to a standard range such as 0 to 1 or -1 to 1. For text datasets, normalization might include converting all texts to lowercase or removing special characters.

The next stage involves splitting your dataset into training, validation, and test sets. This approach helps you evaluate the performance of your model accurately. The training set is used to train the model, while the validation set helps tune it. The test set, which the model hasn't seen before, provides a measure of how well it can generalize to new data. A common practice is to allocate 70-80% of the data for training, 10-20% for validation, and the remaining 10-20% for testing. This division helps ensure that your model is not only well-trained but also well-tested.

For textual data, additional preprocessing steps are needed. Tokenization is the process of breaking down text into smaller units like words or subwords. Lemmatization and stemming can then be applied to reduce words to their root forms, simplifying the vocabulary without losing meaning. Removing stop words, such as "and", "but", and "the", helps reduce noise and focus on the more critical parts of the text. All these preprocessing steps are aimed at making the textual data more digestible for the model.

In cases where you need a custom dataset that is hard to source, consider generating synthetic data. For example, using tools like StyleGAN, you can create new images by training a GAN on a smaller, curated set. This synthesized data can then supplement your primary dataset, making it larger and more diverse. However, be cautious about over-relying on synthetic data, as it might introduce its own set of biases.

Organizing your dataset is the final yet pivotal step before feeding it into your model. Ensure that your data is well-structured, with clear directory hierarchies and proper labeling. For images, this might mean organizing files into subfolders by category. For text, structured formats like CSV or JSON can help keep the data tidy and accessible. Sound data might be organized by genre or instrument.

Documentation can't be overlooked either. Keeping detailed records of data sources, preprocessing steps, and decisions made during data preparation can help maintain transparency and reproducibility. Plus, it makes it easier to return to your project or share it with others in the future. Use code comments, readme files, and even dedicated documentation tools to ensure that all these details are captured.

Finally, remember that data collection and preparation is often an iterative process. As you progress with your AI art projects, you might need to revisit these steps to refine your dataset further. Maybe you discover that your model isn't performing well on certain types of images or that the text generation needs more varied input. Being flexible and ready to adapt will serve you well as you venture deeper into the world of generative AI art.

By carefully considering each aspect of data collection and preparation, you set the stage for successful AI art projects. With a clean, well-organized, and ethical dataset, your generative models are more likely to yield impressive and creative results. Embrace this foundational work, and you'll find it immensely rewarding when you see the stunning and original art your AI can produce.

Ethical Considerations

When diving into the world of AI-generated art, ethical considerations play a pivotal role. As AI technologies continue to evolve and permeate creative fields, it's crucial to examine the ethical landscape that accompanies this transformation. Policies, guidelines, and societal norms are often challenged by the rapid advancements in AI, and this section aims to shed light on key ethical concerns that arise with the use of data for AI art.

One of the most pressing ethical issues pertains to data privacy. AI models used for creating art often require vast amounts of data, frequently collected from the internet. This data can include personal information, images, and other forms of media. The legality and morality of scraping such data without explicit consent come into question. Creators and developers must ask themselves if they're respecting the privacy rights of individuals whose data gets utilized in training AI models. Transparency in data collection practices and obtaining proper consent are fundamental steps to ensure ethical compliance.

Equity and representation are also significant concerns. The datasets used to train AI models often reflect the biases of the society they originate from. If a dataset lacks diversity, the resulting AI-generated art could perpetuate stereotypes or exclude certain groups altogether. For instance, if the majority of data features Western art styles and subjects, the AI's output might overlook non-Western forms and perspectives. Ensuring diverse and representative data could help alleviate this bias, contributing to more inclusive and balanced AI-generated art.

Another critical aspect is the potential for AI art to infringe on intellectual property rights. Many AI models are trained on existing works of art, often without the original creators' permission. This raises questions about the ownership of AI-generated art. Does the output belong to the creator of the AI, the person who generates the artwork, or the countless artists whose work was used to train the model? Clear guidelines and legal frameworks are necessary to address these ownership issues, protecting the rights of original artists while encouraging innovation.

The environmental impact of large-scale AI operations is an emerging ethical issue that cannot be ignored. Training sophisticated AI models requires substantial computational resources, leading to significant energy consumption and carbon emissions. As the world grapples with climate change, it's important for developers and artists to consider the environmental footprint of their AI projects. Embracing energy-efficient algorithms and sustainable computing practices could be a step toward mitigating this impact.

Transparency and explainability in AI art are also vital. Users and audiences should understand how AI models arrive at their creative outputs. While deep learning models often function as "black boxes," there is a growing demand for explainable AI. This transparency can foster trust and allow users to make informed decisions about the ethicality and authenticity of the AI-generated art. It encourages creators to develop models that are not only powerful but also interpretable.

Ethical considerations also extend to the potential misuse of AI-generated art. Deepfake technology, for example, has highlighted how generative AI can be weaponized to create misleading or harmful content. AI-generated art could be used maliciously to forge artwork, create counterfeit pieces, or spread misinformation. Developers and artists need to be vigilant about these risks and adopt measures to prevent such misuse. Implementing watermarking techniques or other forms of digital authentication could help in maintaining the integrity of AI-generated works.

Accessibility is another important ethical factor. AI tools and technologies for art creation should be accessible to a wide range of people, not just those with technical expertise or substantial financial resources. Democratizing access to AI art tools can enable a more diverse group of creators to participate in the AI art movement, fostering innovation and encouraging different perspectives. Open-source tools and educational resources can play a significant role in making AI art creation more inclusive.

Finally, the impact on human creativity deserves consideration. While AI can augment human creativity, there are concerns about it potentially diminishing the value of human-made art. By inundating the market with AI-generated works, we might undermine the appreciation for the unique and irreplaceable qualities of human craftsmanship. It's essential to strike a balance where AI complements rather than competes with human creativity, valuing both forms of artistic expression.

In conclusion, ethical considerations in the realm of AI-generated art are multifaceted and complex. They span issues of privacy, representation, intellectual property, environmental impact, transparency, misuse, accessibility, and the preservation of human creativity. As AI continues to revolutionize the art world, addressing these ethical concerns with thoughtful, well-informed approaches will be key to harnessing its full potential in a responsible manner.

Chapter 9:
Creating Visual Art with
Generative AI

Bridging the realms of technology and aesthetics, creating visual art with generative AI opens up a world where algorithms and creativity coalesce in remarkable ways. By leveraging various AI models, artists can generate unique, captivating pieces that push the boundaries of conventional art forms. Whether it's training a neural network to mimic the styles of famous painters or utilizing GANs to dream up surreal landscapes, the potential for innovation is vast. Enthusiasts can start with accessible tools and gradually delve into more complex techniques, finding freedom in the blend of human intuition and machine precision. As you embark on this journey, remember that every pixel generated is a step toward discovering new modes of artistic expression, challenging the traditional paradigms, and expanding the vistas of visual creativity.

Basics of AI Art

Diving into the basics of AI art means understanding how machine learning models, particularly generative algorithms, can create visual pieces of work. AI art involves the use of algorithms to generate images, videos, and other visual media in a way that often mimics human creativity. Although it's a technological field, the outcome can be profoundly artistic and evocative. With the right tools and understanding, just about anyone can create compelling AI art.

Central to creating AI art is the concept of using an algorithm that learns from a dataset. This dataset typically consists of images that the algorithm analyzes in order to understand patterns, styles, and elements of visual composition. Through learning, the algorithm becomes capable of generating new images that are both original and inspired by the data it has been trained on. This is where the magic begins—a machine learning process producing something new and unexpected.

One of the foundational techniques in AI art is the use of Generative Adversarial Networks (GANs). GANs work by pitting two neural networks against each other: one generates the art (the generator), while the other tries to detect if the art is real or generated (the discriminator). Over time, the generator improves its craft, producing increasingly sophisticated images. This competitive element drives the generation of high-quality, intricate visuals. However, GANs have their complexities and are just one way to approach AI art.

Another important concept is that of Variational Autoencoders (VAEs). VAEs also focus on generating new data that is similar to a given dataset but they do so differently than GANs. VAEs try to compress the data into a smaller, latent space and then decode it back into an image. This technique allows for more control over the generated images, as one can interpolate between points in the latent space to get variations of generated images. Both GANs and VAEs form a good chunk of the AI artist's toolkit, each offering unique capabilities and advantages.

Understanding these models isn't just about the math and algorithms behind them; it's also about grasping the artistic potential they offer. When you train a GAN or a VAE on a dataset of classical paintings, for instance, the output can evoke traditional aesthetics with a modern twist. The results can be both unanticipated and delightfully novel. This convergence of technology and art pushes the boundaries of what we consider creative expression.

Tools and libraries like TensorFlow, PyTorch, and RunwayML have made it increasingly accessible for enthusiasts and beginners to start experimenting with generative AI. These platforms offer pre-built models and user-friendly interfaces that allow you to focus more on the creative aspects rather than the technical hurdles. Such tools serve as great entry points, helping you to understand the basics without needing a deep dive into complex code.

However, knowing the tools is only half the battle. The other half is understanding how to prepare your data and set your goals. A poorly chosen dataset can yield unimpressive or even nonsensical art, while a well-curated one can produce visually stunning results. Therefore, putting thought into what data you use and how you preprocess it is crucial. Whether you're pulling images from museum collections or your own photography, preparation can significantly influence the final outcome.

Creativity in AI art also comes from tweaking and experimenting. Small changes in the dataset, model parameters, or even in the structure of the neural network can lead to startlingly different outcomes. It's a practice of iteration, refinement, and sometimes happy accidents. As with any form of art, there are no fixed rules, and the more you experiment, the more you discover new dimensions of creativity.

In addition to technical skills and creativity, there's also an emotional and philosophical aspect to creating AI art. Questions often arise: What does it mean for a machine to create art? How does this new form of art fit within the broader context of human artistic endeavors? These are questions worth pondering as you dive into the world of AI-generated visuals. Far from being purely academic, these questions can inform your practice, guiding you toward more thoughtful and reflective creations.

The beauty of AI art lies in its blend of predictability and surprise. Algorithms can generate images based on learned data patterns, yet the specifics of each generated piece are often unpredictable. This generates a delightful tension between what you expect the machine to produce and what it actually does. In this interplay, you'll find moments of serendipity, where the machine's 'mistakes' or 'choices' lead to something truly unique and compelling.

As you continue to explore and experiment, you'll find that AI art isn't just about creating; it's also about collaborating. You and the algorithm form a partnership, with each iteration of work building upon the last. This ongoing feedback loop of creation and refinement can lead to increasingly sophisticated and interesting pieces. In essence, you're not just a creator using a tool—you become a collaborator, working in tandem with an intelligent system.

Finally, as you master the basics of AI art, remember that it's a burgeoning field. New techniques, tools, and algorithms are constantly being developed. Keeping up-to-date with the latest advancements will not only enhance your technical skills but also spur your creative imagination. Engaging with online communities, attending workshops, and continually experimenting will keep you at the forefront of this exciting intersection of technology and art.

In summary, the basics of AI art lay the groundwork for an exploration filled with both technical learning and creative discovery. From understanding algorithms to preparing datasets, to using tools and tweaking parameters, the journey is as enriching as the destination. So jump in, experiment, and let the confluence of art and artificial intelligence unfold in your unique creations.

Tools and Techniques

Embarking on the journey of creating visual art with generative AI seamlessly blends imagination with groundbreaking technology. The tools and techniques you select can make all the difference. Just as a painter chooses brushes and a sculptor their chisels, you too must align your creative vision with the most suitable AI tools and techniques. This section delves into the array of digital instruments and methods that will become your co-creators.

Let's start with software tools. There are several popular libraries and frameworks that can be instrumental in generating visual art. *TensorFlow* and *PyTorch* are the go-to options for many machine learning enthusiasts. These libraries offer pre-built models and extensive documentation. Though they may seem complex at first, a bit of practice will reveal their true potential and flexibility. Another widely-used tool is **Processing**, an open-source programming language and IDE built for electronic arts and new media projects. Its versatility makes it well-suited for leveraging generative models in visual art.

Then there's *Runway*, a high-level platform tailored for artists and designers who want to use AI without delving too deep into coding. It provides a user-friendly interface for various generative models, making sophisticated AI accessible to creatives with limited technical skills. With Runway, you can experiment with pre-trained models or even train your own, opening up endless possibilities.

An essential concept worth grasping is the role of **frameworks**. Frameworks bridge the gap between your high-level artistic ideas and the intricate workings of machine learning algorithms. One such framework is *Processing.py*, a Python implementation of the Processing software that makes it easier to incorporate AI. Another option is *Fast.ai*, which offers out-of-the-box solutions with simpler APIs, making it user-friendly for those new to the field. These frameworks act as scaffolding, enabling your creativity to soar without requiring an in-depth understanding of every technical detail.

For those interested in collaboration and sharing their creations, platforms like *GitHub* and *Colab* are indispensable. GitHub is a repository hosting service where you can share code, get feedback, and even collaborate with other artists and developers. Colab, short for Google Colaboratory, provides a cloud-based Jupyter notebook environment where you can write and execute Python code. It's particularly useful for projects that require significant computational power, as you can leverage Google's GPUs and TPUs for faster processing times.

Data is the lifeblood of any AI project, and this holds true for generating visual art as well. **Data collection** and **preparation** are crucial steps that shouldn't be glossed over. High-quality, diverse datasets lead to more nuanced and compelling art. Websites like *Kaggle* and *Google Dataset Search* offer a treasure trove of datasets covering a wide range of topics and styles. Once you've sourced your data, cleaning and pre-processing it becomes the next challenge. Tools like **OpenCV** and **Pandas** can aid in this process, ensuring your data is in optimal shape for training models.

Another key aspect is choosing the **right model architecture**. Convolutional neural networks (CNNs) are often employed due to their ability to handle image data effectively. Variational autoencoders (VAEs) and generative adversarial networks (GANs) are popular choices for generating new images based on input data. Each architecture has its strengths and weaknesses, and the choice depends on the type of art you intend to create. For example, GANs are particularly adept at generating realistic images, whereas VAEs are known for their ability to encapsulate a broad range of styles within their latent spaces.

The training phase itself is an intricate dance. You'll often have to fine-tune hyperparameters such as learning rate, batch size, and the number of epochs to achieve desirable results. It's a process that involves much trial and error, but the reward is well worth the effort. Tools like *TensorBoard* can help monitor and visualize the training progress, offering insights into how your model is evolving and where adjustments are needed.

Post-training, the fun begins. Techniques such as **style transfer** can transform mundane images into stunning works of art. Style transfer involves taking the style of one image and applying it to the content of another. Tools like *DeepDream* and **Neural Style Transfer** frameworks make this process more accessible, enabling intriguing results with relatively minimal effort.

Fine-tuning doesn't stop at the technical aspects. **Human-AI collaboration** plays a significant role in the creative process. Once the models generate output, your artistic interpretation comes into play. Whether it's curating the best results, tweaking the generated images, or combining multiple outputs, your input refines the final piece. Platforms such as *Adobe Photoshop* and *GIMP* can further enhance and polish the AI-created art with your unique touch.

Don't overlook the significance of **feedback loops**. Incorporating iterative feedback can lead to substantial improvements in your work. Use social media platforms, online art communities, and AI forums to gather critiques and suggestions. Engaging with these communities not only provides valuable insights but also fosters collaboration and inspiration.

Finally, keep an eye on emerging **techniques and trends**. The field of generative AI is ever-evolving, with new models and methodologies continually enriching the landscape. Staying updated with the latest research and developments can provide fresh ideas and innovative ways to enhance your art. Websites like *ArXiv* and AI-focused blogs are excellent resources for staying informed about the latest breakthroughs.

In summary, the tools and techniques for creating visual art with generative AI are diverse and powerful. By selecting the right combination of software, frameworks, and methods, you can bring your unique artistic vision to life. The key is to experiment, collaborate, and continually refine your approach. The fusion of human creativity and AI capabilities promises a future where art knows no bounds.

CHAPTER 10:
GENERATING MUSIC WITH AI

Imagine merging the soul of a human composer with the computational prowess of artificial intelligence. This chapter explores the thrilling frontiers of generating music with AI, where algorithms don't just follow notes but create symphonies. We'll delve into how AI models, armed with deep learning and neural networks, can both mimic and innovate upon traditional musical structures. These AI systems can be trained on vast datasets of classical, jazz, and contemporary music, then generate novel compositions that range from hauntingly beautiful to intriguingly experimental. With accessible tools and software, such as OpenAI's MuseNet or Google's Magenta, even enthusiasts can experiment with crafting unique melodies. As we tear down the barriers between technology and artistry, you'll discover that AI-generated music is not just a futuristic concept but a burgeoning reality, ripe for exploration and innovation.

Introduction to AI Music

Welcome to the fascinating world of AI Music, a domain where artificial intelligence and creativity blend to produce melodies and harmonies that can astound the senses. In our journey through "Generating Music with AI," the first step is understanding what AI Music entails and realizing its potential and applications. This venture begins with a look at how AI systems, employing various algorithms and models, can learn, interpret, and create music by mimicking the way humans understand and generate musical compositions.

AI Artistry

Artificial Intelligence, particularly Generative AI, has significantly impacted many creative fields, and music is no exception. By leveraging advanced machine learning techniques and neural networks, AI systems can compose music that rivals—or even complements—what human artists can produce. But what makes AI music truly revolutionary is its ability to draw from vast datasets, break down complex musical structures, and generate fresh, original compositions at an unprecedented scale. This section explores the foundational elements that make this possible, setting the stage for deeper dives into the tools and techniques covered in later chapters.

The origins of AI in music trace back to early experiments with algorithmic composition. Early pioneers employed simple rules and probability-based systems to generate music, but today's AI systems have evolved far beyond these rudimentary methods. Modern AI algorithms can analyze and learn from existing compositions, understand various musical styles and genres, and even generate sheet music that professionals can perform. This combination of machine learning and artistic creation has opened the door to new possibilities, enabling both amateur musicians and seasoned composers to explore uncharted musical territories.

Understanding AI music involves first recognizing the core components of music itself: melody, harmony, rhythm, and timbre. These elements are the building blocks that AI systems must learn to manipulate effectively. For instance, neural networks can be trained to recognize patterns in melodies and harmonies, enabling them to generate sequences that are both coherent and musically appealing. The nuances of rhythm and timbre add another layer of complexity, requiring sophisticated models to capture the subtleties of tempo and instrumental textures.

Deep learning and neural networks, particularly Convolutional Neural Networks (CNNs) and Recurrent Neural Networks (RNNs), play a critical role in the AI music generation process. CNNs are adept at recognizing patterns in large sets of data, making them suitable for analyzing musical scores and audio tracks. RNNs, with their ability to handle sequences, excel in generating new sequences of notes and rhythms that maintain a logical flow. These techniques enable AI systems to "understand" music on a deeper level, facilitating the creation of compositions that can evoke emotion and tell a story.

One might wonder, "Can AI truly replicate the creativity inherently present in human composers?" While AI lacks the emotional experiences and subjective interpretations that human artists bring to their work, it can nonetheless produce compositions that are surprisingly creative and moving. By learning from a diverse range of musical styles and historical contexts, AI systems are uniquely positioned to blend genres, innovate new sounds, and even collaborate with human artists to create unique hybrid compositions. The synergy between human creativity and artificial intelligence pushes the boundaries of musical exploration, leading to innovative and experimental pieces that would be nearly impossible to conceive in isolation.

It's important to highlight the collaborative nature of AI music generation. AI tools are not here to replace human musicians but to augment their capabilities. For instance, AI can assist in generating background scores, suggesting chord progressions, or even creating entire symphonies based on a few user-defined parameters. This empowers musicians to focus on higher-level creative decisions while the AI manages repetitive and time-consuming tasks. The result is a more efficient workflow, enabling artists to produce higher-quality music faster.

Moreover, AI music technology has democratized the music production process. Today, one doesn't need to be a seasoned musician or have access to a professional studio to create compelling music. With AI-powered tools, enthusiasts and beginners can compose, edit, and produce music right from their computers or even mobile devices. This democratization fosters a more inclusive and diverse musical landscape where people from all walks of life can express their creativity.

The applications of AI in music go beyond composition and production. In music education, AI tools can provide personalized tutoring, assist with practice routines, and offer real-time feedback on performance. These intelligent systems can adapt to the learning pace of students, making music education more accessible and effective. For instance, AI-driven apps can analyze a student's playing, identify mistakes, and suggest corrections, providing a highly personalized and interactive learning experience.

The concert experience is another area where AI has begun to make its mark. Live performances are increasingly incorporating AI-generated visuals and sounds to create immersive experiences. Some artists use AI to generate live music in real time, reacting to audience feedback and environmental factors, thereby creating a unique performance for each show. This intersection of AI and live music performance opens new avenues for audience engagement and artistic expression.

While the potential of AI in music is vast, it's not without its challenges. One significant concern is the question of originality and authorship. When an AI generates a piece of music, who owns the copyright? Can AI-created music be considered truly original if it heavily relies on existing compositions for training? These questions don't have straightforward answers and continue to spark debate among musicians, technologists, and legal experts. The ethical implications, which we'll explore in depth in later chapters, are crucial for shaping the future landscape of AI music.

In conclusion, the introduction to AI music serves as a primer for understanding the fundamental principles and incredible possibilities this technology brings to the musical domain. By delving into how AI systems can learn from, interpret, and generate music, we hope to inspire you to explore AI's creative applications further. Whether you're a budding musician looking to experiment with new tools or an enthusiast curious about the intersection of technology and art, AI music offers a rich and exciting field to explore.

Tools and Techniques

Generating music with AI is an exhilarating intersection of technology and artistic creativity. But before diving in, it's essential to understand the tools and techniques that can help bring your musical ideas to life. By leveraging the right tools, you can navigate the complexities of AI-driven music composition and create something truly unique. Whether you are a beginner or an enthusiast, grasping the essentials will pave the way for your exploration. This section will guide you through the most effective tools and techniques for generating music with AI, equipping you with the knowledge to start your own projects.

One of the first steps is to choose the right software tool. There are several popular tools available, each with its own strengths and weaknesses. Some of the most widely-used include Magenta Studio by Google AI, Amper Music, AIVA, and OpenAI's MuseNet. Magenta Studio offers a comprehensive suite of plugins designed for music creation, with a focus on blending human artistry and machine learning. Amper Music and AIVA are excellent for those looking to create royalty-free music efficiently, leveraging pre-configured models to streamline the process. MuseNet, meanwhile, stands out for its ability to generate music with intricate structures, spanning a variety of genres and styles.

After selecting your software, the next step involves setting up your workspace. To ensure smooth operations, it's crucial to have a powerful computer with ample processing power and memory. Music generation can be computationally intensive, and a dedicated hardware setup can significantly enhance your workflow. Additionally, make sure to install all necessary libraries and dependencies. Many AI music tools rely on packages such as TensorFlow, PyTorch, and various audio processing libraries. Having these pre-installed can save you valuable time down the line.

The core of AI music generation often lies in the machine learning models employed. These models are trained on extensive datasets containing various genres, styles, and compositions. One commonly used technique involves using neural networks, particularly Recurrent Neural Networks (RNNs) and Long Short-Term Memory (LSTM) networks. These are designed to handle sequential data, making them well-suited for music, which is inherently sequential in nature. By training these models on large datasets, they can learn intricate patterns and structures, allowing them to generate coherent and creative musical pieces.

Another exciting avenue to explore is Generative Adversarial Networks (GANs). Originally popularized in image generation, GANs are increasingly being applied to music. The basic idea involves two neural networks: a generator and a discriminator. The generator attempts to create plausible music samples, while the discriminator evaluates their authenticity. Over time, this adversarial relationship pushes the generator to produce increasingly sophisticated and realistic music. GANs can be particularly useful for creating new, experimental sounds that push the boundaries of conventional music theory.

For those keen on more structured and orchestrated compositions, Variational Autoencoders (VAEs) offer another robust technique. VAEs excel in generating structured data, making them suitable for composing intricate musical pieces with detailed arrangements. By transforming input data into a latent space and then reconstructing it, VAEs can introduce variations and nuances that add depth to the generated music. This technique is particularly useful for genres that require a high level of complexity, such as classical music.

Moreover, Markov Chains present a simpler but effective method for music generation. They operate on the principle of predicting the next note or chord based on the previous sequence. While not as sophisticated as neural networks or GANs, Markov Chains can still generate surprisingly cohesive and appealing music, especially for genres that rely on repetitive patterns, like techno or ambient music.

Data preparation is another critical component. The quality and diversity of your training data can make or break your project. Collecting high-quality music samples that span different genres, tempos, and instruments will provide your model with a rich source of inspiration. Pre-processing steps such as normalization, segmentation, and augmentation help ensure that your data is in the best possible state for training. This might involve converting audio files into MIDI format, trimming silence, or normalizing volume levels.

When it comes to fine-tuning your model, hyperparameter optimization is key. Adjusting parameters like learning rate, batch size, and the number of layers can dramatically impact the quality of the generated music. Tools like TensorBoard can aid in visualizing and tracking your model's performance, making it easier to identify which parameters yield the best results. Experimentation and iteration are vital—don't be discouraged if the first few attempts don't meet your expectations. With time and tweaking, the model's output will improve.

Beyond the technical setup, a significant part of generating music with AI involves creativity and experimentation. Don't hesitate to incorporate unconventional sounds or techniques. For instance, layering AI-generated tracks with live instruments or vocals can produce a hybrid genre that blends the best of both worlds. You can also experiment with different input formats, such as starting with a simple melody and letting the AI build upon it, or using AI to generate chord progressions as a foundation.

Collaborative creativity can also be a game-changer. Sharing your generated pieces with a community of like-minded individuals can provide valuable feedback and new perspectives. Platforms like AI Music Creativity on Reddit or open-source communities on GitHub offer spaces for sharing, critique, and collaboration. Engaging with these communities can fuel your inspiration and expose you to cutting-edge techniques and innovations in AI music generation.

Finally, evaluate and refine your work continuously. Use both quantitative metrics, such as loss functions and accuracy rates, and qualitative assessments, like musicality and emotional impact, to measure the success of your projects. Combining these approaches will give you a well-rounded understanding of your model's performance and areas for improvement. Keep iterating and refining, embedding new learnings and ideas into your practice.

In conclusion, generating music with AI is a dynamic and rewarding endeavor that merges the realms of technology and artistry. By leveraging the right tools and techniques, from neural networks and GANs to VAEs and Markov Chains, you can unlock new dimensions of creativity. Set up your workspace effectively, prepare high-quality data, and engage in ongoing experimentation and community collaboration. Armed with these strategies, you're well on your way to creating mesmerizing musical compositions powered by AI.

CHAPTER 11:
TEXT GENERATION AND NLP

Venturing into the realm of Text Generation and Natural Language Processing (NLP) opens up an expansive vista where artificial intelligence seamlessly intertwines with human language, enabling an array of creative and practical applications. NLP serves as the backbone of modern text generation, allowing machines to understand, interpret, and even generate human language with uncanny precision. While mastering the intricacies of linguistic nuances might seem daunting at first, the potential for AI to aid in creative writing, from brainstorming story ideas to crafting full-fledged narratives, provides a compelling motive to delve deeper. This chapter will shed light on the fundamental principles driving NLP, examining how algorithms dissect and emulate human language, and highlighting the transformative impact of AI on creative writing. Jumping into the practical aspects, expect to uncover the tools and techniques that bridge the gap between human creativity and machine-generated text, marking a pivotal chapter in your journey through generative AI.

Fundamentals of NLP

Natural Language Processing (NLP) forms the bedrock of many generative AI applications, particularly those related to text. By definition, NLP is a subfield of artificial intelligence focused on the interaction between computers and human language. This involves enabling machines to understand, interpret, and generate text in a way that is meaningful and useful. The journey into the fundamentals of NLP begins with understanding its core components and methodologies.

One of the most crucial aspects of NLP is text preprocessing. Before any advanced analysis or generation can occur, text data must be cleaned and structured. This includes processes like tokenization, where sentences are broken down into individual words or phrases; stemming and lemmatization, which reduce words to their root forms; and removing stopwords, which are common words that may not carry significant meaning. Achieving clean and structured data sets the stage for more intricate NLP tasks.

NLP models often rely on embeddings, which are numerical representations of words. Think of embeddings as a way to convert words into vectors, where semantically similar words are mapped to similar points in a high-dimensional space. Word2Vec and GloVe are popular algorithms used to create these embeddings. The advent of the Transformer model has significantly advanced the field, introducing mechanisms like attention that allow models to weigh the importance of different words in a context.

Moving beyond preprocessing, understanding the linguistic structure of text is indispensable. Syntax and semantics are two fundamental aspects here. Syntax pertains to the arrangement of words to create well-formed sentences, while semantics deals with the meaning behind those words. Parsing techniques, both syntactic and semantic, help in breaking down sentences into their grammatical components, thereby aiding the machine in understanding context and intent.

Named Entity Recognition (NER) is another essential component of NLP. NER involves identifying and classifying key elements in text, such as names of people, organizations, locations, and other entities. This is invaluable for various applications, from search engines to business intelligence. State-of-the-art models train on vast amounts of labeled data to achieve high accuracy in recognizing and categorizing these entities.

Next comes sentiment analysis, a fascinating application where the goal is to determine the sentiment or emotional tone behind a piece of text. This can range from identifying simple positive or negative sentiment to more nuanced emotions like joy, anger, or surprise. Sentiment analysis finds applications in customer feedback analysis, social media monitoring, and market research, among others.

Another cornerstone of NLP is language translation, facilitated by machine translation models. These models have evolved dramatically from rule-based systems to statistical approaches, and more recently, to neural networks. Google's Transformer architecture has brought in a new era of highly accurate and fluid translations, underpinning services like Google Translate. The ability to translate languages effectively relies on intricate NLP techniques and extensive training data across multiple languages.

Context is king in NLP. Traditional models struggled with understanding context because they processed text in a linear fashion, limiting their capacity to capture long-range dependencies in language. The introduction of Recurrent Neural Networks (RNNs) and Long Short-Term Memory (LSTM) networks marked a significant improvement, but it was the Transformer model with its attention mechanisms that revolutionized how context is understood, enabling more nuanced and coherent text generation.

Generative Pre-trained Transformers (GPT) like GPT-3 have pushed the boundaries of what's possible in NLP. These models are pre-trained on a vast corpus of text and can perform a myriad of language tasks with impressive proficiency. From generating creative writing to answering questions and even engaging in meaningful dialogue, GPT models illustrate the power of large-scale language models trained with advanced NLP techniques.

NLP also intersects with speech in the form of Automatic Speech Recognition (ASR) and Text-to-Speech (TTS) systems. ASR models convert spoken language into text, facilitating applications like voice assistants and transcription services. TTS, on the other hand, converts text into spoken words, providing capabilities for voice synthesis in various applications. Both these systems employ NLP methodologies to ensure that the speech-to-text and text-to-speech transitions are as accurate and natural as possible.

Of course, NLP's reach extends to creative domains as well. Whether it's generating poetry, composing articles, or even scripting dialogues for video games, the principles of NLP empower machines to produce text that not only makes sense but can also captivate and engage. Artists and writers are increasingly exploring collaborations with AI to push the boundaries of traditional creative processes.

Finally, the ethical considerations surrounding NLP cannot be ignored. Issues such as bias in language models, the potential for misuse in generating misleading information, and concerns over privacy are areas that require diligent attention. Addressing these ethical challenges is paramount to ensuring that NLP technologies are developed and deployed responsibly.

In summary, the fundamentals of NLP encompass a range of techniques and concepts that are essential for understanding and working with natural language in computational systems. Whether you're gearing up to build your first text-generation model or simply curious about how machines interpret language, mastering the basics of NLP opens the door to a world brimming with possibilities.

AI in Creative Writing

Artificial Intelligence (AI) has begun to revolutionize many facets of our lives, and creative writing is no exception. By harnessing the power of advanced Text Generation and Natural Language Processing (NLP) techniques, AI can now generate poetry, stories, and even articles that are not only coherent but also emotionally evocative. This section will explore how AI is transforming the realm of creative writing, offering both new tools for authors and unprecedented pathways for storytelling.

In recent years, tools like OpenAI's GPT-3 and other transformer-based models have become incredibly proficient at generating human-like text. These models have been trained on vast swaths of data, encompassing everything from classical literature to modern-day tweets. The result? AI that can mimic a range of writing styles and tones, often with astounding accuracy. Authors can use these tools for anything from brainstorming plots to polishing dialogue.

What's particularly exciting is the potential for collaboration between human writers and AI. Picture this: you're stuck in a plot hole or grappling with writer's block. Rather than staring at a blinking cursor, you can input a few prompts, and the AI generates potential ideas or even entire scenes for consideration. This can make the creative process more fluid and dynamic, turning writing from a solitary task into a collaborative effort.

The flexibility of AI in creative writing is stunning. Some platforms enable writers to fine-tune the generated text according to specific guidelines, such as maintaining a consistent tone or adhering to certain genres. For instance, you might ask an AI to craft a science fiction narrative while keeping elements of romance and mystery intact. This opens up new avenues for experimentation and innovation, making it easier for writers to explore unfamiliar genres.

However, this isn't to say AI-generated text is flawless or infallible. There are instances where the output might feel a tad generic or lack emotional depth. Moreover, AI can sometimes produce text that's repetitive or nonsensical—part of the reason why human oversight is essential. But as developers continue to refine these models, the gap between human and AI writing is steadily narrowing.

Imagine being able to generate background lore for an expansive fantasy universe within minutes or creating interactive stories where the narrative adapts based on readers' choices. These possibilities aren't just theoretical—they're practically within reach, thanks to strides in AI-driven text generation. As these tools become more accessible, writers from all backgrounds can leverage technology to elevate their storytelling.

In educational settings, AI can also serve as a valuable teaching aid. By generating writing prompts or even full-length essays, students can learn about different writing styles, narrative structures, and thematic elements in a more interactive manner. Teachers can use AI-generated text as a starting point for discussions, enabling students to dissect the compositional choices and identify areas for improvement.

Beyond the classroom and the literature world, AI is making its mark in journalism and content creation. Reporters can use AI to quickly draft articles, summarize lengthy reports, or even generate multiple versions of a story to target different demographics. Content creators can rely on AI for generating blog posts, video scripts, and social media content, thereby streamlining their creative workflow.

While the convenience and efficiency offered by AI are undeniable, there's also a growing conversation about the ethical implications of its use in creative writing. Issues such as authorship, originality, and intellectual property rights are becoming more prominent. Who owns the text generated by an AI model? How do we ensure that these tools aren't used to create misleading or harmful content? These are questions that both developers and users need to address as AI continues to evolve.

Moreover, AI's ability to mimic human writing raises concerns about the devaluation of human skill and labor. If AI can produce text that's indistinguishable from that written by a human, what does this mean for the future of writers and content creators? While the technology offers incredible benefits, it's crucial to strike a balance—leveraging AI's capabilities without undermining the value of human creativity.

Despite these challenges, the overall landscape of AI in creative writing is one of opportunity and excitement. As AI continues to improve, it's likely we will see even more sophisticated tools that can understand and generate text with greater nuance and emotional intelligence. This could lead to the creation of entirely new literary forms and genres, as well as more personalized and immersive reading experiences.

For instance, consider the potential of AI in crafting narratives for video games or virtual reality experiences. These platforms demand a high degree of interactivity and adaptability—qualities that AI can excel in. By integrating AI-driven text generation, developers could create storylines that adapt in real-time to players' actions, offering a truly unique experience for each player.

Looking ahead, it's clear that AI will continue to play a significant role in the world of creative writing. The tools and techniques at our disposal today are just the beginning, and the future promises even more groundbreaking advancements. From helping novice writers find their voice to enabling seasoned authors to push the boundaries of their craft, AI is set to be a powerful ally in the journey of storytelling.

In conclusion, the intersection of AI and creative writing is an exhilarating frontier that combines the best of human ingenuity with cutting-edge technology. As AI continues to evolve, embracing its potential while remaining mindful of its limitations and ethical considerations will be key to unlocking its full capabilities. With the right balance, the future of creative writing looks brighter—and more innovative—than ever before.

CHAPTER 12:
AI IN GAME DESIGN

AI has dramatically transformed game design, opening up new horizons for creativity and player engagement. One of the most exciting applications is procedural content generation (PCG), where algorithms autonomously create vast and varied game worlds, enhancing replayability and reducing development time. AI-powered game elements such as adaptive enemies or intelligent NPCs (non-player characters) bring a nuanced dynamism to gameplay, making interactions richer and more immersive. By learning from players' behaviors, these AI systems can tailor experiences that evolve in real-time, offering personalized challenges and narratives. Whether through fascinating quests or lifelike virtual opponents, AI's role in game design is pushing the boundaries of what's possible, allowing developers to craft experiences once thought to be the stuff of dreams. Integrating AI not only revolutionizes how games are made but also fundamentally changes how they're played, ushering in an era of endless possibilities and innovation.

Procedural Content Generation

Procedural content generation (PCG) is an exciting area within AI in game design that focuses on creating game content automatically using algorithms and computational processes. This technique can produce vast amounts of diverse and complex game elements, from landscapes and storylines to characters and items, without requiring manual input from developers. For beginners and enthusiasts, understanding PCG opens up a world of creative possibilities and saves a significant amount of time and resources in game development.

At its core, PCG relies on algorithms to generate content based on a set of predefined rules or parameters. These algorithms can range from simple randomization techniques to more sophisticated methods that incorporate machine learning or neural networks. The level of control over the generated content can vary, allowing developers to specify broad guidelines or fine-tune specific details. This flexibility makes PCG particularly valuable in creating dynamic gaming experiences that feel both unique and coherent.

One of the most common applications of PCG is in the generation of game environments. Imagine exploring an open-world game where each player's experience is different because the landscapes, cities, and dungeons are uniquely generated. This not only enhances replayability but also keeps players engaged. Early examples of PCG in games include *Rogue* and *SPelunky*, which used procedural algorithms to create dungeon layouts that were different each time the game was played.

Another fascinating application of PCG is in the creation of characters and creatures. By using procedural methods, developers can generate a vast array of characters with unique appearances, behaviors, and abilities. This is particularly useful in games with large populations of NPCs (non-player characters), as it helps avoid the monotonous repetition of preset character models. Games like *No Man's Sky* have used these techniques to offer a universe teeming with diverse life forms and ecosystems.

Procedural generation also extends to narrative and quests. Algorithms can craft intricate storylines and missions that adapt to the player's actions and decisions. This dynamic storytelling approach ensures that each playthrough offers fresh experiences and surprises. The game *AI Dungeon*, for example, utilizes deep learning models to generate text-based adventures where the narrative evolves based on player inputs.

Delving into the technical aspects, procedural generation often begins with the creation of a seed value. This seed value is a starting point for the algorithm and determines the outcome of the generated content. By changing the seed, developers can produce entirely different results while using the same algorithm, providing an efficient way to create variability without needing multiple algorithms.

Noise functions are a crucial tool in PCG, used to create natural-looking patterns and textures. Perlin noise and Simplex noise are popular noise functions that generate smooth, continuous surfaces, making them ideal for generating terrains and other organic shapes. These functions produce patterns that mimic the randomness found in nature, offering a more realistic and visually appealing outcome.

Advancements in machine learning have further enhanced procedural content generation. Neural networks, particularly Generative Adversarial Networks (GANs) and Variational Autoencoders (VAEs), can be trained to generate high-quality game assets such as textures, sprite animations, and even entire levels. By learning from a dataset of existing game content, these models can produce new assets that blend seamlessly with the game's aesthetic.

The integration of PCG and AI has led to innovative hybrid approaches that combine the strengths of both techniques. For instance, reinforcement learning can be used to improve the quality of generated content by allowing the algorithm to learn from player interactions and feedback. This iterative process helps refine the content to better match player preferences and enhance the gaming experience.

Despite its many advantages, procedural content generation also presents challenges. One of the primary concerns is ensuring that the generated content maintains a high level of quality and coherence.

While algorithms can produce a vast amount of content quickly, not all of it may be suitable for the game. Developers need to implement robust validation processes to filter out subpar content and ensure consistency.

Another challenge is balancing randomness with meaningful design. Complete randomization can lead to incoherent or unbalanced game elements, while overly restrictive algorithms may stifle creativity and variety. Striking the right balance requires a careful consideration of design principles and player expectations, which can be achieved through iterative testing and player feedback.

The future of procedural content generation looks promising, with ongoing research and technological advancements pushing the boundaries of what is possible. Emerging techniques such as procedural storytelling and adaptive content generation are set to revolutionize how games are designed and experienced. As AI continues to evolve, we can anticipate even more sophisticated and creative applications of procedural content generation in game design.

In conclusion, procedural content generation is a powerful tool in the arsenal of game developers, offering endless possibilities for creating rich, diverse, and engaging game worlds. By leveraging algorithms and AI, developers can automate the creation of game content, saving time and resources while delivering unique experiences to players. Whether you're a beginner or a seasoned enthusiast, exploring PCG can inspire new creative avenues and enhance your understanding of game design.

AI-Powered Game Elements

In recent years, the role of AI in game design has evolved from a mere novelty to a foundational element that shapes the very nature of the gaming experience. AI-powered game elements encompass a broad range of features that enhance gameplay, making it more dynamic, engaging, and unpredictable. From intelligent NPCs (non-playable characters) to adaptive difficulty levels, AI is revolutionizing how games are designed, played, and perceived.

One of the most impactful uses of AI in game design is procedural content generation (PCG). This method involves using algorithms to create game content dynamically, eliminating the need for designers to manually craft each element. By leveraging AI, developers can generate vast, complex worlds that are unique to each player's experience. Consider the procedural generation in titles like "Minecraft" and "No Man's Sky," where entire landscapes and ecosystems are created on the fly. This not only saves development time but also offers infinite replayability and exploration for players.

AI's role in developing NPC behavior and intelligence cannot be understated. Traditional game design employed fixed scripts for NPCs, leading to predictable and sometimes stale interactions. Modern AI techniques enable NPCs to learn from player behavior and adapt their responses accordingly. This leads to more complex, life-like interactions, making the game world feel more immersive. Imagine a scenario where NPCs remember past encounters with the player, altering their behavior based on previous actions. This dynamic adaptation creates a richer narrative and adds depth to the player's journey.

Furthermore, AI can be used to tailor game difficulty in real-time. Adaptive AI algorithms analyze a player's skill level and adjust challenges to match, ensuring a balanced and engaging experience. For instance, if a player is struggling with a particular level, the game might ease the difficulty slightly, allowing them to progress without frustration. Conversely, if a player is breezing through, the AI can introduce tougher challenges to keep the gameplay exciting. This level of personalization creates a more inclusive gaming environment, catering to novices and experts alike.

Beyond NPC behavior and difficulty scaling, AI also plays a crucial role in story generation. Narrative-driven games benefit immensely from AI's ability to generate plot points, dialogue, and character interactions on the fly. Using natural language processing (NLP) techniques, AI can craft stories that branch out in numerous directions based on player choices. The result is a non-linear narrative experience where each decision impacts the game world's unfolding story. Games like "AI Dungeon" exemplify this, offering virtually limitless storytelling possibilities driven by player input.

In combat-focused games, AI-driven opponents provide a significant upgrade over traditional enemy design. These adversaries can employ tactics and strategies that challenge even the most seasoned players. For example, in a first-person shooter game, AI enemies might adapt their tactics based on the player's actions, using cover more effectively or setting ambushes. Such intelligent behavior not only increases the game's difficulty but also its realism, creating a more immersive and satisfying experience for players.

However, the use of AI in game elements isn't limited to complex mechanics alone. Even more straightforward, casual games can benefit from AI. For example, in puzzle games, AI can generate levels that are perfectly suited to the player's current mastery, keeping the game just challenging enough to stay interesting. In mobile gaming, AI can tailor ads and in-game offers based on player behavior, improving engagement and monetization.

Another fascinating application of AI in game design is in social dynamics and multiplayer environments. AI can be used to match players of similar skill levels in multiplayer games, ensuring fair and competitive matchups. Furthermore, in massive online worlds, AI can manage the game economy, regulating resource availability, and item values to maintain balance. This kind of dynamic adjustment keeps the game world vibrant and fair for all participants.

AI can also assist in bug detection and repair processes. Traditionally, identifying and fixing bugs has been a labor-intensive part of game development. AI-powered tools can scan through millions of lines of code and gameplay scenarios to detect anomalies, often faster and more accurately than human developers. This not only speeds up the development cycle but also ensures a smoother, more polished final product for players.

Looking to the future, the potential for AI in game design is nearly limitless. With the advent of more advanced machine learning algorithms and increased computational power, AI could take on roles we've only dreamed of. Imagine games where every character has their own AI, interacting not just with the player but also with each other in meaningful ways. Envision game worlds that evolve organically based on a plethora of variables, from player actions to changes in the real world.

Furthermore, AI isn't just enhancing traditional gameplay but is also opening up entirely new genres. Games like "Detroit: Become Human" and "The Elder Scrolls V: Skyrim" utilize complex AI to drive storytelling and world-building in ways that were previously unachievable. These AI-driven game elements are turning interactive digital experiences into genuinely emergent narratives, where every player experiences a unique storyline.

AI in game design brings a treasure trove of new opportunities for developers and players alike. For developers, AI offers tools to create richer, more varied content with less manual effort. For players, it means games that can learn, grow, and adapt in fascinating ways, leading to more immersive and personalized experiences. With AI-driven innovation, the line between the virtual and real worlds continues to blur, making the fantastical seem almost within reach.

In summary, AI-powered game elements are transforming the gaming landscape. These technologies provide adaptive difficulty, intelligent NPCs, dynamic storytelling, and so much more. As AI continues to evolve, so too will the potential for more immersive, engaging, and personalized game experiences. The future of gaming not only looks brighter but also more intelligently crafted, thanks to the ever-growing capabilities of AI.

CHAPTER 13:
INTERACTIVE AND IMMERSIVE ART

Interactive and immersive art is where generative AI truly dazzles, bridging the realms of technology and creativity in ways that captivate and engage. By leveraging advances in Virtual and Augmented Reality, artists can create dynamic, responsive environments that adapt in real time to the viewer's actions. Imagine a gallery where artworks evolve based on your gaze or a concert where the visuals sync perfectly with improvised AI-generated music. Interactive installations, blending sensors and machine learning, offer experiences that are both participatory and deeply personal. These installations can react to human presence and gestures, creating a symbiotic relationship between the audience and the art itself. The creative possibilities are boundless, inviting beginners and enthusiasts alike to explore, innovate, and redefine the boundaries of artistic expression through generative AI.

Virtual and Augmented Reality

Virtual and augmented reality (VR and AR) are transforming how we engage with art, turning passive observation into immersive experience. The potential of these technologies to reshape the landscape of interactive and immersive art is both exciting and boundless. By leveraging generative artificial intelligence, artists can create worlds that not only captivate the senses but also respond to the participant's actions and preferences.

VR takes you into entirely digital realms, offering an escape from reality into environments that feel remarkably real. Imagine stepping into a painting—where each brush stroke comes alive, creating an evolving landscape as you move through it. This is the magic that VR brings to the art world. AI-driven VR experiences can adapt to the viewer's movements, offering a level of interactivity that traditional art simply can't match. For instance, upon entering a virtual art gallery, the surrounding artwork may morph and shift based on the viewer's gaze, breathing new life into static pieces.

In contrast, AR overlays digital elements onto the real world, enhancing our existing environment with supplementary information or artistic flair. This technology blends the physical and digital art spaces, allowing artists to create interactive installations that merge seamlessly with real-world objects. For example, a sculpture in a park could be augmented via a smartphone app to display moving digital elements, creating a hybrid piece that stimulates both the mind and the senses.

Generative AI is the powerhouse behind many of these innovative VR and AR applications. By utilizing machine learning algorithms, artists can generate unique, ever-changing environments tailored to each user. These AI models can analyze user preferences and behaviors to create personalized experiences, ensuring that no two interactions are ever the same. Imagine walking through a virtual forest where the flora and fauna evolve based on your choices and interactions, creating a bespoke environment in real-time.

Furthermore, the use of generative AI in VR and AR extends beyond visual stimuli. Soundscapes generated by AI can adapt to the viewer's environment and actions, providing an immersive auditory experience that complements the visual elements. This multi-sensory approach can make the experience more cohesive and emotionally engaging. The possibilities are endless when combining VR, AR, and generative AI, offering not just art but an experiential narrative that evolves with each interaction.

One of the fascinating aspects of employing AI in VR and AR is its ability to democratize art creation. Artists with limited technical expertise can use AI tools to design complex, interactive installations without needing to write extensive code. These tools often come with user-friendly interfaces, making them accessible to a broader range of creatives. This democratization fosters a more inclusive art community, promoting diverse perspectives and novel ideas.

The educational potential of VR and AR in the art world is another compelling aspect. Museums and galleries can employ these technologies to create interactive exhibits that teach the history and techniques of various art forms in engaging ways. Students can don VR headsets to explore famous works of art up close, examining details that are not visible to the naked eye. Similarly, AR can bring textbook illustrations to life, offering students a more engaging way to learn about art history and techniques.

While VR and AR offer enormous potential, they also come with their unique set of challenges. Technical limitations, such as the need for high-quality hardware and software, can be a barrier to widespread adoption. Ensuring a seamless and lag-free experience requires significant computational power, which can be costly. Moreover, designing for VR and AR demands careful consideration of user comfort—prolonged exposure can sometimes cause discomfort or disorientation. Addressing these challenges is crucial for the sustained success and evolution of VR and AR in interactive art.

Ethical considerations are equally important when integrating AI into VR and AR art experiences. AI models are trained on vast datasets that can include biases, inadvertently perpetuating stereotypes or cultural insensitivities. Artists and developers must be vigilant about the data they use and the potential biases it may harbor. Transparency in

AI processes and a commitment to ethical guidelines can help mitigate these risks, ensuring that the technology is used responsibly.

Looking towards the future, the synergy between VR, AR, and generative AI is poised to revolutionize not just art but various fields such as entertainment, education, and even mental health. Therapeutic applications are already emerging, utilizing VR environments to treat conditions like anxiety and PTSD. The immersive nature of VR, coupled with AI's ability to tailor experiences, offers powerful tools for both artistic expression and practical applications.

Incorporating AI into VR and AR art also opens up avenues for collaborations between artists from different disciplines. Visual artists, musicians, writers, and tech developers can work together to create multi-faceted experiences that envelop participants in a rich tapestry of stimuli. These collaborative projects can result in groundbreaking works that push the boundaries of what art can be, blending multiple mediums into a single, cohesive experience.

The concept of interactive and immersive art isn't new, but what VR and AR bring to the table is unprecedented. As technology continues to evolve, so will the possibilities for creating ever more engaging and personalized art experiences. AI will undoubtedly play a crucial role in this evolution, offering tools and capabilities that were once the realm of science fiction.

In conclusion, the integration of VR and AR with generative AI represents a monumental leap in the realm of interactive and immersive art. This fusion not only enhances the viewer's experience but also empowers artists to push their creative boundaries like never before. As we continue to explore this exciting frontier, the line between the real and the digital will blur, giving rise to a new era of artistic expression that is as intimate as it is expansive. The future of art is not just something we can look at, but something we can step into, interact with, and even shape ourselves.

interactive installations

Interactive installations stand at the crossroads of art and technology, creating dynamic environments where human interaction is essential for the full realization of the artwork. These installations can be seen as playgrounds where both artists and participants become co-creators of the experience unfolding in front of them. Leveraging generative AI, interactive installations have taken on new dimensions, inviting audiences to engage in unique, multifaceted ways that were previously unimaginable.

One of the key distinctions of interactive installations is their reliance on real-time data and user input. Imagine walking into a room where the environment morphs and adapts based on your movements, the sounds you make, or even the biometric data captured from your body. Such reactive technology adds layers of immersion that traditional art forms lack. This evolving art form turns the static spectator into an active participant, providing a deeply personal and communal experience simultaneously.

Generative AI adds another layer of complexity to interactive installations. Algorithms can generate infinite variations, ensuring that no two interactions are ever quite the same. By incorporating neural networks, artists can design installations that learn and evolve from user interactions. The installation becomes a living entity, constantly adapting and changing its manifestations based on the unique and collective inputs from its audience. This helps to blur the lines between the creator and the creation, as each visitor leaves a unique imprint on the piece.

The technology behind these installations often incorporates elements like sensors, cameras, and microphones, coupled with machine learning algorithms that process and respond to this data in real-time. For instance, an installation might use facial recognition software to adapt visual elements depending on the emotions it perceives in the audience members. Or, it might adapt the soundscape based on the decibel levels in the room, creating a dynamic orchestra directed by the audience itself.

Moreover, the incorporation of virtual reality (VR) and augmented reality (AR) into interactive installations further transforms the audience's experience. Imagine wearing a VR headset and walking through a digital art gallery that responds to your gaze, your gestures, or even your heartbeat. Augmented reality can bring these experiences into the real world, layering digital elements onto physical spaces and allowing users to interact with both realms simultaneously. This fusion of the digital and physical worlds creates an engaging, multi-sensory experience that can evoke a broader range of emotions and thoughts.

One well-known interactive installation that utilizes generative AI is "Tree," an immersive virtual reality experience that transforms participants into a rainforest tree. Created by New Reality Company, "Tree" uses user interactions to influence the growth and experience of the tree. As participants move and gesture, the virtual environment responds in kind, offering a deeply personal connection to nature and raising awareness about environmental issues. This type of installation exemplifies how generative AI can create not just art, but impactful messages and experiences.

In the realm of public installations, generative AI can transform urban environments into living canvases. Imagine a city square where the pavement lights up in intricate patterns as people walk across it, or a building facade that changes its appearance based on the weather or the time of day. These installations can make art a part of daily life, enriching the urban landscape and inviting constant interaction from the public.

The possibilities for interactive installations are virtually limitless when you consider the rapid advancements in AI and machine learning. Artists can experiment with combining generative AI with other mediums such as haptic feedback, which provides users with tactile responses, or biofeedback, where the installation reacts to physiological signals like heart rate and brain activity. Such innovations are pushing the boundaries of how we define art and the roles of the artist and audience.

Of course, the creation of these installations comes with its own set of challenges. Technical expertise is required to integrate AI systems seamlessly with other hardware and software components. Artists must also be mindful of ethical considerations, particularly when collecting and processing data from participants. Consent and privacy issues need to be addressed to ensure that interactions remain respectful and secure.

Moreover, from a creative standpoint, the challenge lies in striking a balance between control and randomness. Artists must decide how much influence they want to exert over the resultant experience versus how much they wish to leave to chance and user interaction. This balancing act is crucial for creating installations that are engaging yet coherent, allowing for both unpredictability and meaningful patterns to emerge.

Interactive installations using generative AI also have significant educational potentials. Museums and educational institutions can harness these technologies to create engaging, interactive exhibits that make learning a hands-on, immersive experience. For instance, in a science museum, an interactive exhibit could use generative AI to simulate ecological systems, allowing visitors to see the immediate consequences of actions like deforestation in a highly visual and impactful way.

Furthermore, these installations can play a crucial role in social and behavioral research. By analyzing how users interact with the installation, researchers can gain insights into human behavior, preferences, and social dynamics. Such installations can act as living laboratories where data is constantly being generated, providing valuable information for further studies in both art and science.

The future of interactive installations looks incredibly promising, with continuous advancements in AI, machine learning, and related technologies. As these tools become more accessible, artists from various backgrounds will have the opportunity to experiment with creating immersive environments that defy conventional artistic boundaries. Generative AI opens new avenues for creativity, allowing for the creation of personalized, evolving, and responsive artworks that challenge our perceptions of both art and technology.

In conclusion, interactive installations leveraging generative AI represent a thrilling frontier in the landscape of contemporary art. They invite audiences to become part of the art-making process, transforming passive observation into active participation. This art form not only showcases the incredible potential of technology but also redefines the relationship between the artwork, the artist, and the audience. As we continue to explore the capabilities of generative AI, interactive installations will undoubtedly play a significant role in shaping the future of immersive and interactive art.

CHAPTER 14:
ETHICAL AND SOCIETAL IMPLICATIONS

The rise of generative AI not only opens up remarkable creative possibilities but also poses significant ethical and societal challenges that we must thoughtfully address. As AI systems increasingly contribute to art, questions about originality, ownership, and copyright become more complex. Who owns the rights to a piece created by an algorithm? Moreover, the use of AI in generating content raises concerns about authenticity and potential misinformation. Beyond legal and intellectual property issues, there are broader societal implications to consider. For instance, the accessibility and democratization of art creation through AI tools must be weighed against the risk of eroding traditional artistic skills and livelihoods. It's crucial to foster a balanced approach where innovation thrives, but ethical boundaries are respected, ensuring that AI serves as a tool for augmenting human creativity rather than replacing it.

AI Ethics in Art

When discussing AI ethics in art, we're delving into a complex web of moral questions, cultural considerations, and social impacts. The intersection of artificial intelligence and artistic creation prompts us to examine our definitions of authorship, creativity, and even humanity itself. Ethical matters in this field don't just surface at the moment of creation; they span the entire lifecycle of the AI art process—from data collection to the end user's experience.

One of the primary ethical concerns is the nature of data used to train these AI models. Most generative AI systems require extensive datasets, often sourced from existing artworks, music, or literature. This leads to questions about consent and the rights of original creators. If you're using data scraped from the internet without permission, are you inadvertently misappropriating someone else's intellectual property? This issue is particularly fraught in the art world, where originality and ownership are highly prized.

Moreover, the way these AI models are trained can introduce and amplify biases. If the dataset is not diverse or inclusive, the AI's output will reflect these limitations. For instance, an AI trained predominantly on Western art may fail to represent the rich tapestry of global artistic traditions. This lack of representation perpetuates a skewed view of art and culture.

Another dimension of ethical concern is the potential for AI-generated art to be used in manipulative or malicious ways. Deepfakes, which use advanced generative models to create eerily realistic videos, are a stark example of how AI can be weaponized. While deepfakes can have legitimate uses in art and entertainment, they also pose significant risks to privacy and authenticity. When art can be so convincingly falsified, the stakes for trust and verification rise dramatically.

On the flip side, AI in art also opens doors for unprecedented creativity and democratization. Tools that enable anyone to create art with the help of AI can lower barriers to creative expression. This can lead to a more inclusive art world where people who might not have traditional artistic skills or training can still produce meaningful work. However, this raises the ethical question of whether widespread access diminishes the value of traditional artistry.

The commercialization of AI-generated art introduces yet another ethical layer. When AI-generated art is bought and sold, who ought to be credited as the creator—the human who programmed the AI, the

AI itself, or perhaps even the dataset subjects who inadvertently contributed to the final product? The question of authorship directly impacts how artworks are valued and who benefits financially from their sale.

Furthermore, the environmental impact of AI can't be ignored. Training large generative models requires significant computational resources, which in turn consume considerable energy. The carbon footprint of developing AI technologies is an ethical issue in its own right, raising concerns about sustainability and the broader implications for our planet.

At the intersection of art and technology, we find exciting opportunities for collaboration but also a need for ethical vigilance. Efforts to create guidelines and standards for ethical AI development in the arts are ongoing. These include ensuring diversity in training data, obtaining consent from data contributors, and being transparent about the AI's role in the creation process.

Many in the AI and art communities advocate for a balanced approach where human oversight and machine capabilities complement each other. They argue that rather than seeing AI as a competitor, we should view it as a co-creator. This perspective encourages collaborative ethics, where both human and machine contributions are acknowledged and respected.

In the end, the ethical landscape of AI in art is as dynamic and evolving as the technology itself. As we continue to push the boundaries of what's possible, it's crucial that we remain anchored in ethical principles. This will ensure that AI becomes a tool for positive transformation rather than a force for unintended harm.

Ownership and Originality

Generative AI has revolutionized the art world, but with it come complex questions about ownership and originality. When an algorithm produces an artwork, who owns the rights? The programmer who designed the algorithm? The user who input the parameters? Or maybe even the machine itself, in a figurative sense? These are not just philosophical musings; they are pressing issues with real-world implications in intellectual property law and artistic integrity.

Origination is traditionally tied to the individual artist and their unique expression. However, generative AI blurs these lines. When AI creates, it synthesizes from vast datasets, mixing and reinterpreting countless sources. This raises ethical questions: Is the AI truly creating something new? Or is it merely repackaging existing works? This is especially relevant in cases where the AI's generated output closely mimics the input data, making it difficult to ascertain if the creation is genuinely novel or derivative.

Consider an AI-generated painting. The algorithm that made it was trained using thousands of images. It's conceivable that subtle nuances from different artworks could influence the final output. This convolution can make it challenging to pinpoint the true origin of the piece, muddling the concept of originality. On one hand, one might argue that the AI-generated work is a new, original creation. On the other, it's hard to ignore the fact that the work is heavily influenced by pre-existing art, raising concerns about unintentional plagiarism and copyright infringement.

Another layer to consider is the role of the human operators. In the generative AI equation, humans are both creators and curators. They select the training data, tweak the algorithms, and fine-tune the parameters that ultimately shape the output. Does this make them the rightful owners of the final piece? Or are they merely facilitating the creative process of the AI? This hybridization of roles complicates traditional notions of authorship. The legal ramifications are similarly complex, with courts and lawmakers struggling to keep pace with technological advancements.

An important facet of this debate centers around the concept of the "creative spark". Historically, creativity and inspiration have been viewed as inherently human traits. In the context of AI, some argue that creativity is not genuine unless there's a human element of inspiration, spontaneity, or emotional depth. Others counter that while AI may not experience emotions, its ability to produce unique, compelling works should qualify as a form of creativity, albeit a different one.

Monetization further complicates ownership issues. Artists rely on selling their works to sustain their livelihoods, and AI-generated art is increasingly becoming a part of the commercial art world. When a generative AI artwork fetches a significant sum at auction, the question of who profits becomes pressing. Should it be the developer of the AI? The person who operated it? Or maybe both? Collaborative ownership models are emerging, but they are far from standardized, leaving many grey areas.

The integration of AI in art also impacts collective creativity. Open-source generative models enable a broader range of individuals to create sophisticated works without traditional artistic training. This democratization of art can be seen as a positive development, allowing for a more diverse array of voices and perspectives. However, it could also lead to oversaturation of the art market and devaluation of human-generated art.

In legal terms, copyright law is currently struggling to adapt to these new forms of creation. Traditionally, copyright has been granted to human authors for their original works. Some jurisdictions are starting to recognize AI-generated works, but the majority still do not. As AI becomes more omnipresent, there's a growing need to redefine copyright to encompass these new forms of authorship. This might involve adapting existing laws or creating new ones specifically for AI-generated content.

Moreover, the issue of data sovereignty can't be ignored. AI requires vast amounts of data to function, most of which are sourced from publicly available content. This raises ethical considerations about the use of such data. Is it fair to use the collective creative output of society to train models that can then generate profit for individuals or corporations? There's an ongoing discussion about fair use and the need for explicit consent from original creators when their work is used for training AI.

One approach to handling ownership might involve hybrid models of attribution, where both the human contributors and the AI have defined roles in the creative process. For instance, a generative art piece might credit the algorithm designer, the person who set the parameters, and perhaps even the sources of the training data. This multi-layered attribution could help recognize the different contributions and ensure ethical transparency in the creative process.

Furthermore, questions of agency arise. As AI becomes more sophisticated, it also becomes a more active participant in the creative process. While today's AI doesn't possess intent or self-awareness, it's not hard to envision a future where AI might push boundaries in ways that surprise even its creators. In such a scenario, questions about agency and responsibility become even more crucial: Who's to be held accountable for the content produced by AI if it inadvertently causes harm or offends? These are dilemmas that showcase the intricate interplay of ethics, law, and technology.

The creative industries are starting to adapt, experimenting with new business models and legal constructs to accommodate AI-generated works. Artists are exploring collaborative projects where humans and machines co-create, emphasizing the symbiotic relationship rather than a competitive one. Legal frameworks might eventually adapt to include concepts like "AI-assisted artistry" or "collaborative creativity," which acknowledge the shared roles in the creative output.

In conclusion, the integration of generative AI in the art world brings unprecedented opportunities for creativity and innovation. However, it also introduces complex ethical questions and challenges traditional notions of ownership and originality. As AI continues to evolve, it's crucial for legal, ethical, and artistic communities to collaborate in finding solutions that balance innovation with respect for original creators. By doing so, we can navigate this new landscape in a way that honors both human and machine contributions to the rich tapestry of creative expression.

CHAPTER 15:
CASE STUDIES IN AI ART

Exploring the evolution and impact of AI art can be best understood through real-world applications and projects that have pushed the boundaries of creativity. In this chapter, we'll delve into a variety of notable projects and artists who have harnessed the power of generative AI to create compelling and often provocative works of art. Take, for instance, the mesmerizing visuals generated by AI programs trained on thousands of paintings, capturing the essence of human creativity and imagination while adding a distinctly computational twist. We'll also look at collaborative efforts where artists and engineers joined forces, producing hybrid art forms that challenge traditional concepts of authorship and originality. These case studies provide rich insights and lessons learned, revealing both the potential and limitations of AI as a creative tool. Whether it's generating music, visual art, or interactive experiences, the marriage of technology and creativity showcased in these projects underscores the transformative possibilities that generative AI brings to the art world.

Notable Projects and Artists

In recent years, the landscape of art has been dramatically transformed by the advent of generative AI. Among the myriad projects and artists contributing to this revolution, several stand out for their innovation, impact, and the sheer beauty of their work. These pioneers have harnessed the power of AI to push the boundaries of what is possible in art, creating pieces that challenge our perceptions and inspire awe.

One such artist is Mario Klingemann, whose work explores the intersection of human creativity and machine learning. Often referred to as a "neurographer," Klingemann utilizes neural networks to generate unique visual art. His notable project, "Memories of Passersby I," involves a machine that continuously creates an endless stream of portraits. These portraits, though generated by an algorithm, possess a haunting human quality that invites viewers to ponder the nature of identity and memory.

Then there's the acclaimed collective known as Obvious, who made headlines with their AI-generated portrait, "Edmond de Belamy." This piece, created using a Generative Adversarial Network (GAN), was the first AI-generated artwork to be auctioned at Christie's, fetching an astonishing $432,500. The portrait combines classical aesthetics with modern technology, capturing the essence of both in a single image. Obvious has seamlessly integrated AI into the traditional art world, spotlighting the potential of this technology in creating marketable, high-demand pieces.

Another artist who has made significant contributions to AI art is Refik Anadol. He is known for his immersive installations that utilize large-scale data sets and real-time AI processing. Anadol's "Melting Memories," for instance, visualizes data derived from EEG recordings of brain activity associated with memory. The result is a mesmerizing, dynamic sculpture that evolves in real time, creating a profound connection between mind and machine. Anadol's work not only exemplifies the aesthetic possibilities of AI but also serves as a bridge between art and neuroscience.

Holly Herndon, an avant-garde musician, and composer, has incorporated AI into her music to produce groundbreaking auditory experiences. Her album "PROTO" involves an AI "baby" named Spawn, trained to collaborate with human singers. The interaction between human and machine voices in Herndon's compositions creates a futuristic soundscape that challenges traditional notions of music production and performance. Her work highlights how AI can be a collaborator in artistic creation, rather than just a tool.

Furthermore, AI art is making waves in the fashion industry, thanks to designers like Robbie Barrat. Barrat uses GANs to generate new fashion designs that blend classical styles with contemporary flair. His AI-driven approach allows for the rapid creation of innovative concepts that might take human designers much longer to produce. By exploring the potential of AI in fashion, Barrat is helping to redefine the future of clothing design and production.

The realm of AI art is also profoundly influenced by the contributions of Anna Ridler, who merges the worlds of machine learning and hand-drawn art. Ridler's project "Mosaic Virus" uses historical tulip bulb prices to generate a dynamic visual narrative about economic speculation and commodity trading. The project blends data visualization with aesthetic expression, offering not just an art piece but a commentary on economic history. Ridler's work is a testament to the narrative power of AI art, illustrating how data can be transformed into compelling stories.

Without a doubt, another notable mention is the work of Sougwen Chung, who engages in collaborative drawing performances with a robotic arm. The robot, trained on Chung's previous works, assists her in creating new pieces in real-time. This symbiotic relationship between human and machine offers a unique exploration of authorship and creativity. Chung's performances are a live testament to the possibilities of human-AI collaboration in the arts, demonstrating the fluidity between human intuition and machine precision.

The collective efforts of these artists and their projects underscore the diverse applications of AI in art. From visual arts to music and fashion, AI has become a versatile and powerful medium for creative expression. Each project presents a unique perspective on how machines can aid in the creative process, offering new tools and techniques to artists while opening up uncharted territories for artistic exploration.

Moreover, these notable projects and artists serve as inspiration and motivation for both seasoned artists and beginners alike. They highlight the potential of AI to democratize art creation, offering new pathways for those who might have been limited by traditional methods. By studying these pioneering efforts, one can gain invaluable insights into the possibilities of AI in art and be encouraged to experiment with this technology.

Furthermore, artists like Klingemann, Anadol, and Ridler demonstrate that AI art is not just about the final product but also about the process. The blend of algorithmic complexity with human input opens a dialogue about creativity, authorship, and originality. It challenges us to rethink what constitutes art and who—or what—can be considered an artist.

In summary, the extraordinary works by these notable projects and artists are shaping the future of art in profound ways. They not only showcase the power of generative AI to create beautiful and thought-provoking pieces but also push the boundaries of what we consider possible. These artists are the torchbearers of a new artistic movement, one where the lines between human and machine creativity blur, offering endless possibilities for the future.

Insights and Lessons Learned

Exploring case studies in AI art offers a wealth of insights and lessons that can serve both as inspiration and guidance for enthusiasts and beginners alike. By examining the successes, challenges, and creative breakthroughs of various artists and projects, we can uncover valuable principles that inform and elevate our own work.

One significant insight is the importance of iterative experimentation. In the realm of generative AI, the process of trial and error plays a crucial role in achieving compelling results. Artists often begin with broad concepts and, through continuous tweaks and adjustments, refine their models to better align with their visions. This iterative process isn't exclusive to experienced practitioners; even those new to the field can benefit from a willingness to experiment and adapt their approaches based on the outcomes they observe.

Another key lesson is the role of collaboration, both human and machine. AI tools can function as creative partners, offering unexpected possibilities that might not surface through traditional methods. The case studies often highlight moments where an algorithm's output, initially unintended, sparks a new creative direction. This symbiotic relationship underscores the potential for AI to expand our creative horizons, rather than replace human intuition and skill.

Furthermore, data is the bedrock of generative AI. The quality and nature of the input data profoundly influence the output. Case studies reveal that successful projects pay meticulous attention to collecting, curating, and preparing datasets. This includes ethical considerations, ensuring that the data respects privacy and intellectual property rights. Striking a balance between diverse and rich datasets while being ethically responsible is a lesson that can't be overstated.

A recurring theme across successful AI art projects is the importance of understanding the tools and technologies involved. Mastery doesn't necessarily mean being an expert in every nuance, but having a solid grasp of the underlying principles and capabilities of the technologies you're using. This foundational knowledge allows artists to better troubleshoot issues, optimize their workflows, and push the boundaries of what's possible.

The notion of embracing unpredictability is another valuable insight. Generative AI, by its nature, often produces results that are unexpected and sometimes even bizarre. Artists who approach these surprises with an open mind can transform them into unique and compelling pieces. The ability to view these "happy accidents" as opportunities rather than setbacks can lead to innovative and original creations.

Case studies also highlight the power of storytelling through AI art. Beyond technical prowess, some of the most impactful projects weave narratives that resonate with audiences. Whether it's visual art, music, or interactive installations, framing the work within a meaningful context can elevate its impact and engage viewers on a deeper level. Artists who effectively communicate their vision and the story behind their work often find greater success and connection with their audience.

An essential lesson is the necessity of patience and persistence. Developing AI art can be a time-consuming and complex endeavor. The artists featured in case studies often recount numerous challenges and setbacks, but it's their persistence and dedication that lead to breakthrough moments. This perseverance is a key takeaway for anyone embarking on their own AI art journey.

The importance of community cannot be overlooked. Many groundbreaking projects benefit from the collective knowledge and support of the AI art community. Engaging with forums, attending conferences, and collaborating with others can provide fresh perspectives, technical assistance, and inspiration. The case studies highlight instances where community feedback and collaboration have been pivotal in overcoming obstacles and refining artistic visions.

From these case studies, we also learn that AI art is not a solitary pursuit but part of a larger dialogue within the art world and society at large. Engaging with the ethical and societal implications of AI is crucial. This involves not only considering the immediate impact of one's work but also contemplating how AI-generated art contributes to broader conversations about creativity, originality, and technology's role in art.

Innovation often arises from stepping outside one's comfort zone. Many artists profiled in case studies venture into unfamiliar territories, blending different disciplines and technology to create hybrid forms of art. For example, integrating virtual and augmented reality into AI art projects can result in immersive and interactive experiences that push the boundaries of traditional art forms.

Lastly, the evolving landscape of AI art suggests that adaptability is a key skill. Technology and techniques are constantly advancing, and artists who are open to learning and evolving with these changes will be better positioned to take advantage of new opportunities. Staying informed about the latest developments in AI and maintaining an ongoing commitment to growth can help artists remain at the forefront of this dynamic field.

In conclusion, the case studies in AI art provide a rich repository of insights and lessons that can guide and inspire anyone interested in this exciting field. By embracing experimentation, collaboration, and storytelling, understanding the tools and technologies, and engaging with the community and ethical considerations, enthusiasts and beginners alike can unlock the creative potential of generative AI. Patience, persistence, and a willingness to adapt will serve as invaluable assets on this journey, enabling artists to not only create compelling works but also contribute meaningfully to the evolving landscape of AI and art.

CHAPTER 16:
UNDERSTANDING AI ART CRITICISM

AI art criticism is a fascinating field that intersects technology and aesthetics, requiring a nuanced understanding of both. As AI generates art, the question of how to evaluate its quality becomes paramount. Traditional art critics often rely on subjective interpretations, but AI art evaluation introduces metrics like originality, coherence, and technical proficiency. This chapter delves into these evaluation metrics and explores how public perception varies from admiration for AI's creative potential to skepticism about its authenticity. As you dive deeper, remember that the way AI art is critiqued can significantly shape its future development and reception, ultimately defining its place in the broader art world.

Art Evaluation Metrics

When we dive into the world of AI art, the conversation often pivots to evaluating the artistic quality of the creations. This isn't just abstract musing—art evaluation metrics serve as the backbone for determining the success and impact of generative AI art. Understanding these metrics helps us bridge the gap between human perception and machine-generated art.

Traditional art criticism relies heavily on subjective interpretations and personal tastes. However, when it comes to AI-generated art, a blend of subjective and objective metrics often gives a more comprehensive view. The metrics used encompass a variety of factors—ranging from aesthetic quality and creativity to technical execution and the novelty of the piece. Balancing these components is crucial to forming a well-rounded critique.

The aesthetic quality of AI-generated art can be subjective and varies from one viewer to another. To make this evaluation more structured, metrics such as color composition, symmetry, texture coherence, and visual harmony are often employed. These factors are akin to those used in traditional art but are adapted to fully appreciate the nuances of AI creations.

Creativity is another core metric. It gauges how innovative and original a piece of art is. For AI-generated art, this often involves looking at how the AI deviates from known patterns and creates something that stands apart from existing works. Algorithms like Generative Adversarial Networks (GANs) and Variational Autoencoders (VAEs) contribute significantly to this creative aspect. Evaluators seek to understand the extent to which the AI has ventured into new territories versus rehashing established styles.

The technical execution of the artwork is paramount, especially in AI art, where the capacity of algorithms is showcased. Evaluators check for clarity, resolution, and the seamless integration of different elements in the artwork. Art evaluation metrics might also look at the underlying techniques the AI used—such as the complexity of neural networks involved and the sophistication of data preprocessing steps.

Novelty plays an indispensable role. An AI art piece should present something unseen before—offering a fresh perspective or unfamiliar forms. Novelty is often intertwined with cultural and historical contexts, making it a vital but complex metric to assess. In the art community, novelty can sometimes be polarizing; what is seen as groundbreaking by one viewer might be perceived as obscure by another.

To add another layer of assessment, we should also consider audience engagement. How do viewers interact with and respond to the AI-generated artwork? This includes social media shares, gallery exhibit placements, and even viewer comments. Metrics here can be quantifiable, like the number of likes and shares an artwork receives, or qualitative, such as analyzing viewer sentiments expressed in feedback.

Additionally, narrative consistency can be another useful metric. This refers to how well the AI-generated art maintains a cohesive story or theme. Even if the art is abstract, viewers tend to seek out patterns or narratives they can relate to. Evaluating this can involve a mix of both machine-led and human-centered approaches, ensuring the created narrative aligns with broader artistic intent.

A sometimes overlooked, yet crucial metric, is the ethical implications behind the artwork. Ethical evaluations consider the sources of data used for training the AI, the potential biases introduced, and the societal impact of the artworks. If an artwork is technically proficient but based on ethically questionable data, it can lead to significant controversies.

Peer recognition and critique also act as subtle yet profound metrics. In the art world, gaining acknowledgment from other artists and critics can be a testament to the work's quality and impact. For AI art, this means the piece must stand up to the scrutiny of both technologists and traditional artists, creating a unique challenge for evaluative metrics.

Lastly, let's discuss scalability and reproducibility—two metrics particularly relevant for AI art. Scalability refers to how the art can be adapted or expanded upon, either by the same AI, a different AI, or even by human artists collaborating with AI. Reproducibility, on the other hand, assesses if the AI's creative process can be reliably reproduced to achieve similar artistic outcomes. Both metrics offer insights into the potential growth and influence of AI-generated art.

To properly evaluate AI art, it's essential to combine multiple metrics to form a holistic understanding. This composite approach allows us to appreciate the layers of complexity involved in both the creation and the eventual audience reception of these works. Through these lens, we can begin to tease apart the intricate web of technology, creativity, and artistic expression, shedding light on what makes AI-generated art a fascinating and evolving frontier.

Public Perception

Public perception of AI-generated art is a blend of fascination, skepticism, and curiosity. On one hand, the general excitement around the novelty of machines creating art has captured the imaginations of people who might not otherwise engage with technological advancements. The sheer innovation and capabilities of generative AI to produce engaging, sometimes breathtaking pieces, often leave audiences in awe. Some perceive these creations as a glimpse into the future of art, offering a fresh perspective and new possibilities that challenge traditional art ownership and creation.

Conversely, there's a significant portion of the public that remains wary of AI-generated art. Skepticism arises primarily from a perceived lack of authenticity and humanity in the work produced by algorithms and neural networks. Art, traditionally a deeply personal and human endeavor, feels almost sacrosanct to these critics. Therefore, the notion that a machine, devoid of emotional experiences or intentionality, can create meaningful art seems disingenuous to them. For these individuals, AI-generated art may be interesting as a technological experiment, but it falls short of what they consider true artistic expression.

Incredible as it may seem, AI art has even sparked heated debates about creativity and originality. A common question is whether AI can truly be creative or if it merely mimics the patterns found in human-produced art. The public grapples with understanding where the line between replication and innovation lies. These discussions often revolve around definitions of creativity, prompting a reevaluation of what it means to be an artist in the age of intelligent machines.

Moreover, the perception of AI art is influenced by its presentation and context. When showcased in tech conferences or digital forums, the focus tends to be on the underlying technology and innovation. However, when these artworks are displayed in traditional art spaces, such as galleries and museums, they invite a different kind of scrutiny and appreciation. The environment can shape the viewer's interpretation, sometimes granting the work a legitimacy it might not have in a purely technological setting.

Social media also plays a pivotal role in shaping public perception. Platforms like Instagram, Twitter, and TikTok are flooded with AI-generated artworks, making them accessible to a broad audience. However, this immediacy and proliferation can also lead to oversaturation, where the novelty wears off, and the art becomes just another trend. How AI artists and technologists manage their creations in these spaces greatly impacts how enduring or fleeting these works appear to the public.

A particularly thorny issue impacting public perception is the ethical dimension of AI art. Concerns about copyright, artistic ownership, and the potential for job displacement create a complex backdrop that heavily influences how AI art is received. When people see artworks generated by algorithms trained on existing human-made pieces, questions about consent and exploitation arise. These ethical dilemmas can taint the public's view, causing some to reject AI art on principle.

Public perception is not static; it evolves as both the technology and its applications grow. Generative AI is still in its relative infancy, which means that as it matures, public opinion will likely shift. Early detractors may become more accepting as they see advancements that address their concerns, while early adopters will continue to push the boundaries of what AI art can achieve, reshaping the narrative around it.

In educational settings, generative AI art has also begun to gain acceptance. Schools and universities are increasingly incorporating AI art into their curricula, helping students understand not only the technological aspects but also the philosophical and ethical discussions it engenders. This integration helps demystify the technology for younger generations, fostering a more informed public perception rooted in knowledge rather than speculation.

It's also worth noting that public perception varies widely across different cultures and communities. In some regions, technological advancements in art are met with enthusiasm and support, while in others, they may be viewed with suspicion or even hostility. These differing viewpoints can stem from varying levels of technological adoption, cultural attitudes towards art and creativity, and socio-economic factors that influence how new technologies are embraced or resisted.

Artists themselves play an important role in shaping how their AI-generated works are perceived. How they position their work within the broader art community, their willingness to engage in dialogues about their processes and motivations, and their responses to criticism all contribute to the public narrative. By transparently sharing their creative journeys and the role of AI in their work, artists can help demystify the technology and foster a deeper understanding of its creative potential.

The media's portrayal of AI art cannot be understated in shaping public perception. Headlines that either overly hype the capabilities of AI or, conversely, dismiss its potential can skew popular opinion. Balanced reporting that acknowledges both the innovative potentials and legitimate concerns surrounding AI art helps cultivate a more nuanced and informed public perspective.

Chris Elliott

Lastly, the future of public perception regarding generative AI in art will likely be influenced by the level of inclusivity and democratization within the AI art community. If the tools and platforms for creating AI art remain accessible and open to a wide variety of people, the resulting diversity of voices and expressions could help broaden acceptance. However, if these tools become monopolized by a few, there might be backlash against perceived elitism or commercialization, impacting how AI art is viewed by the broader public.

In summary, public perception of AI art is multifaceted and continuously evolving. It's shaped by a complex interplay of excitement, skepticism, ethical considerations, and cultural contexts. As generative AI continues to advance, both its critics and proponents will play crucial roles in defining how this new form of art is understood and appreciated. Through open dialogue, education, and mindful creation and presentation, the public can come to see AI-generated art not just as a novelty but as a meaningful extension of human creativity.

CHAPTER 17:
CONNECTING WITH THE AI ART COMMUNITY

Immersing yourself in the AI art community is one of the most rewarding aspects of exploring generative AI. This vibrant ecosystem is brimming with diverse voices, innovative ideas, and invaluable resources that can greatly enhance your journey. Online forums and groups provide spaces for sharing experiences, seeking advice, and showcasing your work, allowing you to grow alongside fellow enthusiasts and professionals. Participation in conferences and events offers unique opportunities to network, learn from leading experts, and stay updated on the latest trends and advancements. By engaging with this dynamic community, you not only refine your skills and knowledge but also contribute to a collective effort that pushes the boundaries of what AI can achieve in the realm of art.

Online Forums and Groups

One of the most effective ways to immerse yourself in the AI art community is by joining online forums and groups. These platforms not only provide a wealth of resources and information but also offer opportunities to connect with like-minded individuals. From exchanging tips and tricks to getting feedback on your work, online forums can serve as your launchpad into the world of AI-generated art.

The first step is finding the right communities. Websites like Reddit host a variety of subreddits focused on AI art, such as *r/deepdream* and *r/generativeart*. Here, you can find discussions ranging from technical challenges to the ethical considerations of generative art. Reddit is especially useful for newcomers because it has a voting system that highlights popular content and discussions, making it easier to find valuable information quickly.

Another popular platform is Discord, which features real-time chat rooms where you can engage in conversations, participate in collaborations, or even join live coding sessions. Many public Discord servers are available, such as the ones run by various AI art tool developers and communities centered around specific AI software like RunwayML or Artbreeder. These groups are often buzzing with activity, offering a continuous stream of inspiration and support.

For more formal discussions and deeper dives into the technical aspects, consider platforms like Stack Overflow and specialized AI forums. While these might be more intimidating to beginners, they're an excellent source for troubleshooting specific technical issues and enhancing your coding skills. Over time, participating in these discussions can deepen your understanding of machine learning techniques and the algorithms behind your favorite art-generating tools.

Engaging in these online communities can also open up opportunities for collaborative projects. Artists and developers frequently reach out in these forums with collaboration requests, providing a chance to work on more complex and ambitious projects than you might tackle alone. Collaborative efforts not only yield better work but also allow you to learn from others' expertise, bringing a multidisciplinary approach to your own art.

Facebook groups are another venue where AI enthusiasts gather. Although Facebook's algorithm might not be perfect for discovering niche interests compared to Reddit or Discord, it does offer various groups dedicated to AI art. Joining these can provide a more casual and social atmosphere, where you can share your progress, ask for feedback, and participate in community challenges or events.

Then there's Twitter, which might not be a traditional forum but is invaluable for staying updated on the latest trends and developments. Many AI artists, researchers, and organizations are highly active on Twitter. Following them can keep you in the loop on emerging tools, techniques, and events. The platform's hashtag system makes it easy to find posts related to AI art by searching tags like #ganart, #aiart, or #generativeart.

Online forums and groups are not merely places to lurk and consume content; they're arenas for active engagement. By commenting on posts, sharing your own work, and participating in discussions, you'll build a reputation and make connections that can lead to real-world opportunities. Some forums even host regular competitions or themes that challenge members to create art based on specific prompts or rules, which can be a great way to test your skills and gain visibility.

Don't overlook the importance of smaller, niche communities either. Although they might have fewer members, they often provide a more intimate setting where you can form closer relationships and get tailored advice. Websites like DeviantArt or ArtStation, though traditionally focused on digital art, now have sections and tags dedicated to generative art. Engaging with these communities can provide unique perspectives and inspiration you might not find in larger, more generalized groups.

It's also worth mentioning that many online communities host regular virtual events, like webinars, live streams, and Q&A sessions with experts. Attending these can give you first-hand insights into advanced techniques and emerging trends. Plus, they're yet another way to engage with the community actively and stay motivated.

In conclusion, online forums and groups offer an expansive and dynamic resource for anyone interested in AI art. They provide not only technical support and inspiration but also a sense of community and camaraderie. Whether you're a complete beginner or someone looking to deepen your understanding, joining these virtual spaces can exponentially accelerate your growth and enhance your creative journey.

Conferences and Events

Jumping into the AI art realm can be exhilarating, but one of the most impactful ways to truly immerse yourself is by attending conferences and events. These gatherings serve as melting pots of ideas, offering endless opportunities for learning, networking, and inspiration. Whether you're a novice trying to grasp foundational concepts or an enthusiast eager to explore cutting-edge advancements, conferences can provide the intellectual nourishment you crave.

Major AI art conferences typically feature a rich array of activities, including keynote speeches, panel discussions, hands-on workshops, and live demonstrations. These events often attract an eclectic mix of artists, engineers, researchers, and industry leaders, fostering an environment where cross-disciplinary collaboration thrives. For instance, events like NeurIPS (Neural Information Processing Systems) and ICCC (International Conference on Computational Creativity) highlight groundbreaking research in AI and computational creativity. While NeurIPS focuses on the broader AI community, ICCC zooms in more on creativity, offering a perfect nexus for those interested in generative AI art.

Workshops at these conferences offer more intimate settings where you can get hands-on experience with generative AI tools. You'll often find sessions covering a variety of topics—from using GANs to generate unique pieces of artwork to leveraging VAEs for creating intricate designs. The benefit of these practical workshops is undeniable: they allow you to engage directly with the tools and software that you might otherwise only read about or watch through tutorials.

Equally compelling are the opportunities to witness live demonstrations. Imagine watching an AI model create visual art in real-time, tuned and manipulated by experienced artists and engineers before your eyes. These sessions don't just show what's possible; they spark ideas for your own projects. Seeing an algorithm transform a dataset into stunning visuals can be profoundly inspiring, often serving as a catalyst for your own creative explorations.

Another significant advantage of conferences and events is the chance to hear directly from the pioneers in the field. Keynote speakers often include thought leaders who have made trailblazing contributions to AI art. These talks provide insights into the latest research, share experiences, and sometimes even hint at the future directions of the field. Listening to a pioneer share their journey can be both enlightening and highly motivational.

Panel discussions are a staple at most conferences, offering a platform for experts to debate and discuss current trends, ethical considerations, and future possibilities. These sessions can be enlightening, as they provide multiple perspectives on the same topic, often touching upon issues you may not have considered. For instance, you might hear artists and ethicists discuss the implications of AI-generated art on traditional art forms and the concept of originality.

Networking opportunities are perhaps the hidden gems of these events. Casual conversations during coffee breaks, lunchtime discussions, or evening meetups can turn into lasting professional relationships. These interactions can lead to collaborations and new projects, as well as provide a sense of community. Joining forums and groups post-conference can extend these relationships and provide ongoing support and inspiration.

Specialized sessions focusing on the latest tools and software are another highlight. Industry leaders often use these platforms to launch new tools or present updates on existing ones. Being physically present allows you to ask direct questions, get immediate feedback, and even troubleshoot any issues you might be encountering with your own projects.

Many conferences also feature art exhibitions showcasing the latest achievements in AI art. These exhibitions can be a feast for the eyes, displaying the sheer diversity and creativity that generative algorithms can achieve. Apart from the aesthetic pleasure, these showcases provide a deeper understanding of how different techniques and models can be applied creatively. Walking through these exhibitions, you'll find yourself pondering how to push the boundaries of your own work.

Accessibility is also improving with the advent of virtual conferences. Due to constraints like travel costs or time commitments, attending a physical conference might not always be feasible. Virtual events break these barriers, offering a way to engage with the community from the comfort of your home. Although they may lack the tactile experience of live events, they make up for it with recorded sessions, chat forums, and virtual meetups, extending the learning experience.

Choosing which conferences to attend might feel overwhelming given the plethora of options. It's important to align your choice with your specific interests and your current understanding of AI art. For those just starting, conferences that cover a broad spectrum of topics, offering 101 sessions and foundational workshops, might be more beneficial. Conversely, if you've been dabbling in AI art for a while and want to delve deeper into niche areas, specialized conferences focusing on specific types of generative models or applications might be more suitable.

Apart from formal conferences, there are numerous smaller events, meetups, and hackathons. These smaller gatherings provide a more casual, flexible environment where you can experiment, ask questions, and receive feedback in real time. Participating in a hackathon can be particularly rewarding; these events usually center around creating a project within a short timeframe, encouraging rapid learning and hands-on experimentation.

Finally, don't overlook regional events. Local meetups and smaller conferences can be tremendously valuable, offering the chance to connect with people in your geographic area who share your interests. These local connections can lead to in-person collaborations, studio visits, and the development of a supportive local community of AI artists.

In sum, the value of attending AI art conferences and events is multifaceted. From gaining hands-on experience with cutting-edge tools and techniques to making invaluable connections within the community, these events offer a wealth of knowledge and opportunities. They are crucial milestones in your journey, transforming abstract concepts into tangible skills and ideas, paving the way for your growth as an AI artist. So, the next time you hear about an AI art conference or local meetup, seize the opportunity. You never know what sparks of inspiration you might find or who you might meet to pull you further along in your creative journey.

Chapter 18:
Practical Projects for Beginners

Diving into the world of generative AI can be an exhilarating journey, especially when you start with hands-on projects designed for beginners. In this chapter, we'll explore a series of beginner-friendly projects that can help you harness the power of generative AI and unleash your creativity. These practical projects include step-by-step guides that break down complex concepts into manageable tasks, making it easier for you to follow along and build your confidence. Additionally, we've compiled a variety of resources and templates to jump-start your projects, whether you're interested in creating AI-generated art, music, or even text. By working through these examples, you'll not only learn the basics of generative AI but also gain valuable experience that can serve as the foundation for more advanced explorations. So, grab your tools and let's get started on creating something truly extraordinary!

Step-by-Step Guides

Welcome to the heart of bringing your generative AI concepts into reality—this is where theoretical knowledge meets practical application. The step-by-step guides in this chapter aim to be your hands-on companion, providing clear instructions for projects that span various domains of generative AI. Whether you dream of creating a visually stunning piece of digital art, composing a mesmerizing piece of music, or generating engaging text, you'll find this section invaluable.

Starting with simple projects, these guides will gradually increase in complexity, ensuring that you build both confidence and competence. We'll pay attention to key details, explain the rationale behind each step, and offer tips to troubleshoot common issues. The goal is to demystify the process and make generative AI accessible, even if you've never written a line of code before. So, let's dive in and transform those imaginative ideas into real, working projects!

Before we embark on specific projects, it's crucial to ensure you're set up with the right tools and software, as covered in Chapter 7. Having your workspace properly configured with the necessary libraries and frameworks is foundational. Once you're ready, you can proceed to the exciting phase of hands-on creation.

One of the most straightforward yet fulfilling projects you can start with is "Creating a Random Art Generator using Generative Adversarial Networks (GANs)." GANs are a powerful type of neural network designed to generate data indistinguishable from real data. The project's aim is to get you acquainted with GANs in a controlled environment, starting with pre-existing datasets. You'll learn how to preprocess data, train the model, and produce novel images.

Once you're comfortable with visual data, transitioning into "Text Generation with Recurrent Neural Networks (RNNs)" or Transformer models can be a natural progression. This project will introduce you to generating coherent text snippets, whether poetry, short stories, or even code snippets. Here, you'll grasp the intricacies of sequence models and tokenization, which are the building blocks for handling natural language processing (NLP) tasks.

After mastering text and visuals, a fantastic next step is exploring "Music Generation Using Generative AI." This project will enhance your understanding of the unique challenges and opportunities in generating sequential data. Different tools such as OpenAI's MuseNet or Google's Magenta will come into play, showing how AI can compose music in various styles. This guide will walk you through dataset preparation, model selection, and training, culminating in the digital symphonies you've always wanted to create.

Switching gears, consider delving into "Procedural Content Generation for Game Design." If you have an interest in game development, this guide will reveal how generative models can automate the creation of game elements like levels, characters, and storylines. The focus will be on integrating generative models with popular game engines like Unity, ensuring a seamless experience from code to console.

For those with a vivid imagination, "Creating Interactive and Immersive Art Using Generative AI and VR" is an excellent project to tackle. Virtual Reality (VR) and Augmented Reality (AR) push the boundaries of how users interact with digital content. This guide will walk you through creating immersive art installations that respond to user inputs or environmental triggers. The marriage of AI and VR offers limitless possibilities, and this guide aims to help you harness that potential.

In another vein, "Generating Realistic Human Faces and Avatars" could be an intriguing project. Using Variational Autoencoders (VAEs) and GANs, you'll learn to produce highly realistic human faces. Applications for this technology are vast, from creating virtual influencers to populating video games with photorealistic characters. The guide will provide specifics on data collection, ethical considerations, and the technical steps to achieve lifelike avatars.

Of course, no section on practical guides would be complete without touching on "Ethical and Societal Aspects of Generative AI Projects." While it's exhilarating to see your AI creations come to life, ethical considerations should never take a back seat. This segment will include guidelines on identifying and mitigating biases in your data and models, ensuring your generative projects are fair, inclusive, and do no harm. Understanding these implications imbues your work with responsibility and foresight.

One lesser-known but equally fascinating project is "Generating Synthetic Data for Machine Learning." Synthetic data has applications in scenarios where real data is scarce, sensitive, or costly to obtain. This guide will help you understand how to use generative models to create synthetic datasets that mimic real-world data distributions, which can be pivotal in training robust machine learning algorithms.

Another engaging project is "Creating AI-Powered Story Generators." This combines text generation with a layer of decision-making to produce interactive stories that respond to user inputs. Using various NLP techniques and AI models, you'll build a system capable of crafting branching narratives, offering a dynamic and engaging storytelling experience.

Lastly, consider the "Mid-Range Project: AI Style Transfer for Images." Style transfer lets you apply the artistic style of one image to the content of another. Imagine rendering your photographs in the style of Van Gogh's "Starry Night" or Picasso's "Cubism." This project will cover algorithm selection, implementation tricks, and aesthetic evaluations, culminating in visually stunning results.

Each project includes detailed instructions, code snippets, and extensive troubleshooting tips. As you progress, remember that the scope of these guides is not to make you merely replicate existing projects but to inspire innovation by providing a foundation. The real artistry lies in experimenting beyond these guides to create something uniquely yours.

As you conquer project after project, you'll inevitably discover new challenges and areas for improvement. This iterative learning process, coupled with a curious and open-minded approach, will be your most significant asset. Transform these step-by-step guides into stepping stones for your creative journey in the fascinating world of generative AI.

In conclusion, these step-by-step guides serve as a springboard into the vibrant realm of generative AI. Whether it's visual art, music, text, or game content, each project is designed to impart foundational skills while sparking your creativity. Dive deep, experiment fearlessly, and most importantly, enjoy the process. The world of generative AI is expansive and ever-evolving, and your contributions could be the next big wave of innovation.

Resources and Templates

In the journey to mastering generative AI and creating compelling AI-driven projects, having access to the right resources and templates can be a game-changer. This section aims to provide a comprehensive collection of essential tools, templates, and resources that will cater specifically to beginners embarking on practical projects in generative AI. Finding the right dataset, implementing state-of-the-art models, and fine-tuning your creations require resources that are both accessible and straightforward for newcomers.

Starting with datasets, open-access repositories such as Kaggle, UCI Machine Learning Repository, and OpenAI's datasets are invaluable. These platforms offer a plethora of options ranging from image datasets for visual art to text corpora for natural language processing projects. For instance, Kaggle not only provides datasets but also hosts competitions which often come with robust datasets and exemplary code from top participants. If you're working on a visual art project, you might find ImageNet or COCO (Common Objects in Context) particularly useful.

Diving into pre-trained models, initiatives like TensorFlow Hub and PyTorch Hub offer a vast array of pre-trained models that can kickstart your project. These repositories include models for various tasks such as image generation, language translation, and even more specialized tasks like style transfer. Leveraging these pre-trained models can save significant time and computational resources, allowing you to focus more on the creative aspects of your project.

Code templates and examples are essential when you're just starting. Websites like GitHub and GitLab host thousands of repositories where experienced practitioners share their code, often with detailed readme files and setup instructions. Following these well-documented examples can provide a stepping stone to understanding complex algorithms and methodologies. Additionally, many of these repositories include Jupyter notebooks, which offer an interactive way to explore and tweak the code, making learning more intuitive and less intimidating.

For those interested in specific frameworks, many libraries come with their own documentation and example projects. TensorFlow provides an extensive collection of tutorials and guides that range from beginner to advanced levels. Similarly, PyTorch's tutorials cover everything from the basics to deploying models in production environments. The official documentation and community forums for these libraries are great resources to troubleshoot issues and seek advice from more experienced developers.

An often overlooked but critical component of any generative AI project is the computational infrastructure. Cloud services like Google Colab, which offers free access to GPUs and TPUs, can be incredibly helpful for beginners who may not have high-end hardware. Additionally, platforms like AWS and Azure provide scalable computing resources, allowing you to run complex models and experiments without the need for substantial initial investment in physical hardware.

Pre-built templates can also simplify the initial setup and help you maintain a structured approach to your projects. Templates for specific types of projects—such as creating a GAN for image synthesis or a

VAE for generating novel music—can offer a solid foundation upon which you can build and customize your work. Websites like Papers with Code often link the original research papers with corresponding codebases, giving you access to state-of-the-art methodologies alongside the actual implementation.

Community-driven resources can also be immensely beneficial. Platforms like Stack Overflow, Reddit, and specialized forums such as AI Alignment Forum and GitHub Discussions provide spaces where you can ask questions, share insights, and learn from the experiences of others. These communities are typically very welcoming to beginners, and engaging with them can accelerate your learning process.

For a more guided approach, online courses and tutorials offered by institutions like Coursera, edX, and Udacity are highly recommended. These courses often come with assignments and projects that you can add to your portfolio. Additionally, platforms like Medium and Towards Data Science offer numerous articles and guides written by AI practitioners and researchers, which cover a wide range of topics and practical advice.

Books and scholarly articles are another rich resource. While online articles and tutorials are great for quick learning, diving into textbooks such as "Deep Learning" by Ian Goodfellow, Yoshua Bengio, and Aaron Courville can provide a deeper understanding of the underlying theories and methodologies. For more focused reading, research papers available on arXiv can give you insights into the latest advancements in generative AI.

To further enrich your toolkit, software libraries are indispensable. Python libraries such as NumPy, pandas, and Matplotlib are essential for data manipulation and visualization. For tasks more specific to generative AI, libraries like TensorFlow, PyTorch, and Keras come equipped with numerous functionalities and pre-built models that can significantly speed up your development process. Keras, in particular, is known for its simplicity and ease of use, making it an excellent choice for beginners.

Lastly, don't underestimate the power of tutorials and walkthroughs provided by experienced practitioners. Websites like YouTube and Coursera host numerous video tutorials where experts walk you through complex concepts step-by-step. Channels dedicated to AI and machine learning can be particularly helpful in breaking down sophisticated ideas into more understandable segments.

In summary, the vast array of resources and templates available today can pave the way for a smoother and more engaging entry into the world of generative AI. By utilizing community forums, leveraging pre-trained models, accessing quality datasets, and following structured tutorials, you'll not only save time and effort but also gain a deeper understanding and appreciation of this fascinating field. As you delve into your practical projects, these resources will be your companions, guiding you through the challenges and celebrating the breakthroughs along the way.

CHAPTER 19:
MAINTAINING AND UPDATING YOUR WORK

As you navigate through your journey with generative AI, maintaining and updating your work is key to ensuring longevity and relevance. Version control becomes indispensable in tracking the evolution of your projects, allowing you to reflect on past iterations and learn from your experiments. Continuous improvement, on the other hand, involves a cycle of refining models, incorporating new data, and staying abreast of technological advancements. This ongoing process not only enhances the quality of your creations but also fortifies your understanding and skill set. By embracing these practices, you're not just preserving your work; you're setting the stage for future innovations and discoveries in the vibrant domain of AI-driven artistry.

Version Control

Version control is an essential practice, especially when you're dealing with the dynamic and rapidly evolving field of generative AI. For those new to the concept, think of version control as a sophisticated way of keeping track of changes in your work over time. Whether you're a beginner or an enthusiast, employing version control systems (VCS) can significantly simplify your workflow and enhance your productivity.

Imagine you've spent hours crafting a generative AI model that produces stunning visual art. You decide to tweak a parameter to see if it can make your art even better. A week later, you realize the change wasn't as effective as you'd hoped. Without version control, retracing your steps and reverting to the previous state can be tedious and error-prone. With version control, however, you can effortlessly roll back to that earlier state of the project, saving you both time and frustration.

One of the most significant advantages of using version control in your creative AI projects is the ability to work collaboratively. When multiple people are contributing to a project, it's crucial to have a system that can handle concurrent modifications without overwriting someone else's contributions. Version control systems like Git provide a way to branch out from the main project, allowing team members to work independently on different features before merging their changes back into the main codebase.

By leveraging branching and merging features, version control enables experimentation without risk. You can create a new branch to test out an idea or modify an existing feature. If the experiment is successful, you can merge it into the main project; if not, simply discard the branch. This approach inspires creativity because it removes the fear of making irreversible mistakes.

For beginners, navigating a version control system might seem daunting at first, but the learning curve is well worth it. Tools such as GitHub, GitLab, and Bitbucket offer intuitive interfaces that simplify the process, making it easier to manage your versions effectively. Many of these platforms also integrate with popular Integrated Development Environments (IDEs) such as PyCharm, Visual Studio Code, and Jupyter Notebooks, making them increasingly accessible.

Using commit messages wisely is another crucial aspect of effective version control. When you make a change to your project, commit these changes with descriptive messages that convey what was altered and why. For instance, "Changed learning rate from 0.01 to 0.001 to improve model accuracy" is far more insightful than a generic "Updated parameters" message. Detailed commit messages act as a historical record, allowing you to understand the evolution of your project over time and facilitating easier debugging and collaboration.

Another significant component of version control is branching strategies. Different teams might employ various branching strategies based on their workflow requirements. Some might use a "feature branch" model, where each new feature is developed in its separate branch. Others might use a "release branch," focusing on stabilizing a version before deploying it. Understanding and selecting an appropriate branching strategy can make your project more organized and your workflow smoother.

Labels and tags are additional tools within version control that can help you keep track of significant points in your project's history. Tags can be used to mark releases, making it easier to identify stable versions of your work. For example, you might tag a particular commit as "v1.0" when you are ready to release your first version. This practice is beneficial not only for your reference but also when sharing your work with others, be it collaborators or an audience.

Let's not overlook automated testing and continuous integration (CI) when discussing version control. Automated testing ensures that changes in the code do not break existing functionalities. When coupled with CI services like Travis CI, Jenkins, or GitHub Actions, each change is automatically tested, and you are instantly notified if something goes wrong. This workflow is invaluable in maintaining the integrity of your generative AI projects, especially as they become more complex.

One key to successful version control is consistency. Make it a habit to commit your changes often and employ a descriptive commit message strategy. Regular commits help in tracking your progress and provide a safety net, so you can revert to previous states whenever necessary. It might seem like an overhead initially, but over time, this practice will prove to be a lifesaver.

Imagine exploring different algorithms for generating music with AI. You might have several working versions, and each has its unique twist. Without version control, managing these iterations manually can become unmanageable. However, through version control, each version can reside in its branch, allowing you to switch between them seamlessly while maintaining a clean main codebase. When you find the optimal solution, merging it into the main branch is straightforward and safe.

Documentation is an essential aspect that pairs well with version control. By coupling comprehensive documentation with your versioned projects, you create a robust framework that future-proofs your generative AI endeavors. This way, anytime you return to your project or onboard new collaborators, the documentation serves as a guiding light, ensuring continuity and clarity.

Even in individual projects, version control can be a powerful ally for continuous improvement. By keeping detailed notes and commit messages, you can learn from past mistakes and successes, making your future projects more efficient. Additionally, the discipline instilled by regular use of version control can make you a more meticulous and thoughtful developer.

In the era of open-source contributions, understanding version control is indispensable. Many pioneering generative AI projects and libraries reside in open-source repositories. Contributing to these projects requires familiarity with VCS principles. By engaging with these projects, you not only grow your skillset but also become part of a broader community pushing the boundaries of what's possible with AI.

Embracing version control might appear as an added chore initially, but the long-term benefits are substantial. It equips you with the tools to manage your generative AI projects methodically, mitigate risks, and collaborate effectively. In a field where innovation is rapid and changes are the norm, mastering version control becomes not just an asset but a necessity.

Continuous Improvement

Once you've gotten a grasp of the fundamentals and started creating your own generative AI projects, it's easy to feel a sense of accomplishment. But as with any creative or technical endeavor, there's always room for growth and refinement. Continuous improvement is a journey, not a destination, and it's crucial for evolving your skills and staying up-to-date with the ever-changing landscape of generative AI.

The first step in continuous improvement is to cultivate a mindset of lifelong learning. In the rapidly evolving field of AI, what's cutting-edge today can become outdated tomorrow. Subscribing to newsletters, following influential researchers and practitioners on social media, and joining AI-focused communities can keep you informed about the latest advancements. Participating in online forums or communities, like Reddit's r/MachineLearning or AI art-specific groups, allows you to exchange ideas and get feedback from peers who share your interests.

Another key aspect of continuous improvement is setting specific goals for yourself. Rather than vaguely aiming to "get better," outline concrete objectives you wish to achieve within a set timeframe. For instance, you might want to master a new AI framework, participate in a Kaggle competition, or publish a paper on your findings. Clear goals help you track progress and maintain motivation. Break these goals down into manageable tasks, and celebrate small victories along the way to stay motivated.

Feedback is a crucial component of improvement. It might be tempting to work in isolation, especially if you're a creative type, but feedback from others can offer new perspectives and identify areas for enhancement that you might have overlooked. Participate in hackathons, open mics for AI art, or local meet-ups where you can present your work and receive constructive criticism. This not only helps you improve but also expands your network, offering potential collaborations and new learning opportunities.

Moreover, reflective practice can be incredibly beneficial. After completing a project, take some time to review what went well and what could have been done differently. Maintain a journal where you document your process, challenges, and solutions. This will not only serve as a valuable reference for future projects but also help you recognize patterns in your creative and technical approaches that may need adjustments.

Embracing new tools and techniques is essential for growth. The field of generative AI is awash with innovative tools, libraries, and frameworks. For example, if you're primarily using TensorFlow, you might experiment with PyTorch to see if it offers different advantages for your workflow. Regularly attending workshops and training sessions can help you stay proficient with the latest tools. Many online platforms like Coursera, edX, and Udacity offer specialized courses led by experts in the field.

Revisiting and updating past projects is another effective strategy for continuous improvement. As you learn new techniques, apply them to your older works to see how they can be enhanced. This not only breathes new life into previous creations but also provides a tangible measure of your progress over time. Update your project repositories and documentation to reflect your latest methods and insights. Incorporating version control systems, like Git, ensures that you can neatly organize and manage iterations of your projects.

Peer collaboration can also open new doors for improvement. Working alongside others not only diffuses knowledge but also introduces you to different approaches and ideas. Collaborative projects can challenge and expand your own skill set in ways you might not have anticipated. Participate in group projects, whether they're formal academic endeavors or more casual creative collectives. Learning to merge different styles and techniques is invaluable in the world of AI art, where interdisciplinary approaches often lead to the most innovative work.

Training datasets are the bedrock of generative AI, and their quality significantly impacts the output. Continuously refining your datasets by adding diverse, high-quality data can improve the robustness and creativity of your models. Experimenting with data augmentation techniques or collecting niche datasets tailored to your artistic vision can yield more surprising and engaging results. Ensuring that your datasets are ethically sourced and well-documented not only enhances their quality but also aligns with broader ethical practices in AI.

Staying curious and experimenting are crucial attitudes for continuous improvement. Generative AI thrives on experimentation. Try different model architectures, tweak hyperparameters, or even combine multiple models to see what unique results you can produce. Sometimes, the most unexpected combinations can lead to groundbreaking creations. Maintain a sandbox environment where you can test and iterate without the pressure of finality. This freedom invites creativity and innovation.

The importance of interdisciplinary learning cannot be overstated. Generative AI intersects with various fields such as computer science, art, music, and even psychology. Delving into these other domains can provide fresh inspiration and insights for your AI projects. For example, studying the principles of traditional art can help improve the aesthetic quality of AI-generated visuals, while understanding musical theory can lead to richer and more complex AI-composed music tracks.

Keeping up with academic research can help you push the boundaries of your own work. Accessing academic papers on platforms like arXiv or through university libraries can introduce you to pioneering methodologies and the latest findings in the field. Papers often include detailed breakdowns of experiments, which can serve as blueprints for your own investigations. Attend academic conferences, either in person or virtually, to directly engage with researchers and ask questions that particularize your unique challenges and interests.

Participate in competitions and challenges. Platforms like Kaggle host competitions that can kickstart your learning and offer real-world problems to solve. These competitions often feature complex datasets and specific project requirements, pushing you to apply your skills creatively. Beyond the immediate competition, you can study the winning solutions and methodology to glean new techniques and insights.

Last but not least, it's beneficial to understand and embrace the philosophical and ethical dimensions of your work. Generative AI doesn't just create art; it also raises questions about the nature of creativity, authorship, and even the future role of humans in the creative process. Engaging with these questions can deepen your understanding and appreciation of your work, making it more meaningful and impactful.

By adopting these strategies for continuous improvement, you'll be better equipped to navigate the evolving landscape of generative AI. Your journey will be enriched by new skills, deeper insights, and perhaps most importantly, a vibrant community of fellow enthusiasts and professionals who can share this transformative adventure with you.

CHAPTER 20:
MONETIZING AI ART

As the landscape of creativity evolves, monetizing AI-generated art reveals itself not only as an innovative venture but also a viable income stream for artists and tech enthusiasts alike. By navigating various platforms, you can license and sell your creations, reaching a global audience eager for the next digital masterpiece. Harnessing crowdfunding and sponsorship opportunities further amplifies your reach, offering financial support and validation from a community invested in your success. The possibilities are vast—from offering exclusive prints to creating bespoke pieces for clients—allowing you to turn passion into profit. As you tread this exciting path, remember that the right blend of creativity, business acumen, and technology will empower you to thrive in the burgeoning world of AI art.

Licensing and Sales

Licensing and selling AI-generated artwork can be an exciting yet challenging endeavor. This section will guide you through the essentials of navigating the commercial landscape, offering practical advice on how to monetize your creative outputs while ensuring you respect intellectual property laws and industry standards.

Firstly, let's dive into the concept of licensing. Licensing is essentially the granting of permission to use your artwork under defined conditions. It often involves legal agreements that specify how, where, and for how long the artwork can be used. When it comes to

AI-generated art, it's crucial to understand the different kinds of licenses available, such as exclusive, non-exclusive, and royalty-free licenses.

Exclusive licenses grant the licensee full rights to use the artwork, often restricting the artist from selling or licensing the same piece to others. This type of license can be lucrative but usually comes with a higher price tag. Non-exclusive licenses, on the other hand, allow the artist to license the same piece to multiple buyers. This can be beneficial for reaching a wider audience and generating continuous revenue streams. Royalty-free licenses permit the buyer to use the artwork without paying royalties, but they usually come with more usage limitations compared to other license types.

As a beginner or enthusiast, you might wonder how to start licensing your AI-generated art. One effective way to get your foot in the door is by using online platforms like ArtStation, Adobe Stock, and Shutterstock. These platforms offer exposure to a broad audience of potential buyers who are searching for unique digital art. They also provide built-in licensing frameworks, allowing you to focus more on creating art rather than getting bogged down in legal complexities.

In addition to leveraging online platforms, you may also choose to set up your own website or portfolio. This route gives you greater control over your artwork, pricing, and terms of use. Make sure to include comprehensive information about the types of licenses you offer, along with clear terms and conditions to avoid any misunderstandings down the line.

Another crucial aspect of licensing is understanding the value of your work. AI-generated art is a relatively new field, and pricing can vary widely. Factors such as the uniqueness of the piece, the complexity of the creation process, and the intended use can impact its value. Researching comparable works in the market can provide valuable insights into how to price your art competitively.

Now, let's explore sales. Selling AI-generated art can take many forms, from one-time transactions to ongoing revenue streams. Your approach will depend largely on your goals, your audience, and the nature of your art. Direct sales through your website or digital platforms can be straightforward, allowing buyers to purchase prints, digital downloads, or even physical merchandise featuring your artwork.

Online marketplaces like Etsy, Redbubble, and Society6 offer plug-and-play solutions for selling prints and merchandise. These platforms handle much of the logistics, such as printing, shipping, and customer service, freeing you to focus on your art. It's also worth noting that platforms like these often feature community forums and resources to help you optimize your listings and increase visibility.

Beyond direct sales, consider participating in art shows, galleries, and exhibitions, both online and offline. While these venues are traditionally geared towards conventional artworks, the growing interest in digital art is opening new avenues for AI-generated pieces. Exhibiting your work can not only boost your visibility but also attract potential buyers and collaborators.

For those looking to engage with corporate clients, offering commissioned work can be lucrative. Companies often seek unique and innovative visuals for branding, marketing, and internal communications. AI-generated art can stand out due to its novelty and customizability. To pursue this path, create a compelling portfolio showcasing your capabilities, and don't hesitate to reach out to potential clients directly.

In the current digital age, social media platforms like Instagram, Pinterest, and TikTok can also serve as powerful tools for marketing and selling your AI art. Consistently posting high-quality images, engaging with followers, and collaborating with influencers can drive traffic to your sales channels. Additionally, sponsored posts and targeted ads can help broaden your reach.

While monetizing AI art presents numerous opportunities, it's also important to navigate challenges such as copyright issues and fair use. Generative art often involves substantial training data, which might include copyrighted materials. Ensuring that your data sources are appropriately licensed and credited is vital to avoid legal repercussions. Tools and resources like Google's Open Images Dataset or the Creative Commons can offer alternatives for sourcing legally compliant data.

Transparency with your buyers is another key aspect; whether your art is fully AI-generated or involves human collaboration, clearly communicating the process behind each piece can build trust and authenticity. Providing detailed descriptions, including the algorithms and tools used, can also add an extra layer of intrigue and value for potential buyers.

Finally, think long-term. Establishing a sustainable model for licensing and sales involves continuous adaptation and learning. Staying updated on industry trends, emerging platforms, and evolving legal standards will help you refine your approach and maximize your success. Engaging with the broader AI art community can offer insights, inspiration, and support as you navigate this exciting field.

In summary, monetizing AI-generated art through licensing and sales is multifaceted, requiring a blend of creativity, business acumen, and legal awareness. By leveraging various platforms, understanding your audience, and continuously refining your strategy, you can unlock the full potential of your AI artistry in the commercial realm.

Crowdfunding and Sponsorship

When it comes to monetizing AI art, crowdfunding and sponsorship are two popular avenues worth exploring. Both offer distinct opportunities for creators to secure financial backing while retaining a degree of creative freedom. Let's delve into what each mechanism entails and how you can leverage them to support your generative AI projects.

Crowdfunding is all about rallying a community around your vision. Platforms like Kickstarter, Indiegogo, and Patreon provide creators with a stage to present their projects, garner support, and gain monetary contributions from a broad audience. The allure for backers often lies in the sense of direct engagement and contribution to the development of innovative work.

Creating a successful crowdfunding campaign necessitates meticulous planning. The first critical step is to sculpt a captivating narrative around your AI art project. Explain what makes your work unique, why it matters, and how it pushes the boundaries of generative AI. Visual aids, such as demo videos or sample artworks, can substantially enhance your pitch. They provide potential backers with a tangible sense of what they're supporting.

Another crucial element is setting realistic financial goals. It's essential to calculate your project's expenses accurately, including software, hardware, data acquisition, and personal living expenses during the creation period. Overestimating may deter potential backers, while underestimating could leave you without sufficient resources to complete your project.

Transparency is key to maintaining trust with your backers. Regular updates on your progress, paired with an open dialogue about any challenges or delays, can foster a supportive community around your work. Crowdfunding isn't just about money; it's about building a network of advocates who believe in your vision.

In addition to addressing potential financial backers directly, consider tiered rewards systems to incentivize larger donations. Offering exclusive artwork, behind-the-scenes content, or early access to finished pieces can stimulate higher levels of financial support. Tailor these rewards to reflect the unique aspects of your project and your rapport with your audience.

On the other hand, sponsorship involves forging relationships with companies, organizations, or influential individuals who are willing to support your AI art projects financially. This method often provides more substantial funding compared to crowdfunding, but it may come with specific expectations or conditions regarding the use of funds and the direction of your work.

To attract sponsors, you need to present a compelling case about the mutual benefits of the partnership. Outline how their support can elevate both your project and their brand. Demonstrating a potential return on their investment, whether through brand exposure, technological innovation, or community engagement, can be a significant selling point.

Your professional profile and previous accomplishments play a crucial role in attracting sponsors. Compile a portfolio showcasing your best work, including past projects, exhibitions, publications, and any media coverage that highlights your expertise and achievements in the field of generative AI art. A well-crafted proposal should accompany this portfolio, detailing your project plan, timelines, and budget.

Networking is indispensable for finding potential sponsors. Participate in industry events, art exhibitions, and tech conferences where you can meet representatives from organizations interested in the intersection of art and technology. Formal networking events and casual conversations alike can yield valuable connections.

Building long-lasting relationships with sponsors requires clear communication and integrity. Regularly update your sponsors on the project's progress, share successes and challenges, and acknowledge their contribution publicly whenever appropriate. These practices help solidify trust and pave the way for future collaborations.

Don't overlook the possibility of combining crowdfunding and sponsorship. Initiating a successful crowdfunding campaign can demonstrate to potential sponsors the public interest and initial financial backing for your project. It's like a proof of concept; if you can show that a community is invested in your work, sponsors might be more inclined to contribute substantial resources.

There's also the growing trend of decentralized autonomous organizations (DAOs) in the art world. These entities pool resources from multiple members to support creative projects based on collective decision-making. Exploring this avenue could offer another layer of community-driven support for your work.

While pursuing crowdfunding and sponsorship, be prepared for potential rejection and setbacks. Not every pitch will land, and not every campaign will meet its goal. Consider these experiences as learning opportunities. Analyze what worked and what didn't, adjust your approach, and try again. Persistence and resilience are key attributes for any artist venturing into the realms of generative AI.

Remember, the strategies that work best will often depend on the specifics of your project and your audience. Stay flexible, keep refining your approach, and remain open to new opportunities. Whether through crowdfunding or sponsorship, securing funding for your AI art projects is all about connecting with the right people who share your enthusiasm and vision.

In summary, crowdfunding and sponsorship offer unique, viable paths to monetize your AI art, each with its own set of advantages and challenges. By mastering both, you can enhance your financial stability, broaden your audience, and continue pushing the boundaries of digital creativity.

CHAPTER 21:
FUTURE TRENDS IN GENERATIVE AI

The landscape of generative AI is brimming with promise, poised to transform across numerous domains. Emerging technologies like quantum computing, edge AI, and advanced neural network architectures are pushing the boundaries of what's possible. We're on the brink of creating AI systems that can generate hyper-realistic art, simulate complex environments with astonishing accuracy, and even co-create alongside humans. Predicting future innovations, we can expect a deeper integration of generative AI into everyday tools, making creativity accessible to everyone. This seismic shift hints at not just the evolution of AI technology, but a redefinition of artistic and creative practices themselves.

Emerging Technologies

In the ever-evolving landscape of generative AI, emerging technologies are constantly reshaping what's possible, driving innovation, and expanding the boundaries of creativity. As these technologies develop, they offer new tools and frameworks that enable both novices and experts to explore uncharted territory. This section outlines some of the major emerging technologies, highlighting their potential to revolutionize the field of generative AI.

One of the most exciting frontiers in generative AI is the advent of transformer models. These models, such as GPT-3 and BERT, have demonstrated a remarkable ability to understand and generate human-like text. Unlike traditional neural networks, transformers use self-attention mechanisms that allow them to weigh the importance of different words in a sentence, improving comprehension and generation. This technological leap has profound implications for natural language processing (NLP) applications, from chatbots to creative writing.

Additionally, transformer models are not limited to text. Researchers are actively exploring their capabilities in generating visual art and music. For example, by training transformers on images, we can produce highly detailed and contextually relevant artwork. In music, these models can compose original pieces that replicate various styles or even create entirely new genres. The versatility of transformer models makes them a pivotal element of the future landscape of generative AI.

Another emerging technology deserving attention is neural rendering. This technique leverages neural networks to generate realistic images from 3D models or even from textual descriptions. Imagine describing a fantastical landscape and having a neural network create a photorealistic image of it. Neural rendering is poised to transform industries such as video game design, virtual reality (VR), and augmented reality (AR), enabling the creation of more immersive and interactive experiences.

Further advancements in VAEs (Variational Autoencoders) are also pushing the boundaries of what can be achieved with generative AI. Traditionally used for tasks like image generation and data compression, VAEs are being enhanced with new techniques like hierarchical and disentangled representation learning. These enhancements allow for greater control and specificity in the generated outputs, making VAEs an invaluable tool for artists and designers who require a high degree of customizability in their work.

Flow-based models are another area of innovation. These models offer a different approach to data generation by learning an invertible transformation between a simple distribution and the target data distribution. This property allows for exact sampling and likelihood estimation, making flow-based models highly efficient and flexible. Applications of flow-based models are expanding into areas like real-time image synthesis and anomaly detection, broadening their utility and appeal.

In the realm of hardware, there are significant advancements as well. Quantum computing, although still in its nascent stages, holds immense promise for the future of generative AI. Quantum computers leverage the principles of quantum mechanics to perform complex calculations at unprecedented speeds. As quantum hardware becomes more accessible, it could lead to breakthroughs in optimizing generative models, solving problems that are currently intractable with classical computers.

Edge computing is another hardware-related development to watch. By performing data processing at the edge of the network, closer to the data source, edge computing reduces latency and bandwidth use. This can be particularly beneficial for real-time generative AI applications, such as interactive installations or mobile AI art tools, where quick response times are crucial.

In terms of software, new frameworks and libraries are continually being developed to make generative AI more accessible. Platforms like TensorFlow, PyTorch, and newer entrants such as JAX provide powerful tools for building and deploying generative models. These frameworks are increasingly incorporating features that simplify the training and customization of models, lowering the barrier to entry for beginners and enabling more seasoned practitioners to experiment more freely.

Blockchain technology is also making inroads into generative AI, particularly in the realm of art and IP protection. By creating a decentralized ledger of digital assets, blockchain can ensure the authenticity and ownership of AI-generated art. This is crucial in an age where digital art is easily replicable, ensuring that creators are fairly compensated and their work is protected from unauthorized use.

Moreover, the integration of generative AI with other emerging technologies like IoT (Internet of Things) is opening up new possibilities. For instance, IoT devices equipped with generative AI capabilities can autonomously create tailored content, from personalized music playlists to custom visual displays, enhancing the user experience in smart homes, cars, and other environments.

Generative adversarial networks (GANs) continue to be a hotbed of innovation as well. Researchers are continually discovering new architectures and training techniques to improve GAN performance and stability. Recent developments include techniques to address mode collapse and enhance the quality of generated images. These innovations are making GANs more reliable and effective for a wider range of applications, from creating lifelike avatars to generating high-resolution textures for virtual environments.

Ethical AI is another burgeoning area. As generative AI becomes more sophisticated, questions about ethics and responsible use are taking center stage. Emerging technologies are being developed to ensure that AI systems are fair, transparent, and accountable. Techniques like explainable AI (XAI) are designed to make the decision-making processes of AI models more understandable to humans, contributing to a more ethical deployment of generative AI.

Lastly, collaborative AI platforms are gaining traction. These platforms enable multiple AI systems to work together, learning from each other and pooling their resources to tackle complex tasks. For example, a collaborative platform might integrate text, image, and music generation capabilities to produce multi-modal art installations that could not be created by a single system alone. This collaborative approach not only enhances the creative potential of AI but also fosters interdisciplinary innovation.

The landscape of generative AI is teeming with possibilities, driven by these and other emerging technologies. As these advancements continue, the realm of creative AI will become increasingly rich and diverse, offering endless opportunities for discovery and innovation. Staying abreast of these developments will empower enthusiasts and professionals alike to push the boundaries of what's possible and explore new frontiers in creativity and technology.

In summary, the evolution of generative AI is marked by rapid advancements across multiple fronts. Transformer models, neural rendering, enhanced VAEs, flow-based models, quantum and edge computing, new software frameworks, blockchain, IoT integration, improved GANs, ethical AI, and collaborative AI platforms are just some of the emerging technologies shaping this vibrant field.

These innovations are not only redefining the capabilities of generative AI but also expanding its applicability across various domains. Whether you're a beginner exploring the basics or an experienced creator pushing the limits of your craft, the future of generative AI presents an exciting landscape brimming with potential.

Predicting Future Innovations

The landscape of generative AI is evolving rapidly, offering glimpses into both near and distant futures brimming with potential. As generative models continue to advance, new applications and innovations are continually on the horizon. This section delves into the myriad ways that generative AI may transform society, creativity, and technology in the years to come. What might the future hold for this dynamic field?

One major trend is the integration of generative AI in everyday tools and applications. Imagine a world where generative AI blurs seamlessly with our daily digital interactions. Picture smart personal assistants that don't just respond to your needs but anticipate them by generating personalized content—be it music, artwork, or even tailored news articles. These systems could offer unparalleled personalization, making our digital experiences richer and more meaningful.

Advances in computing power, particularly through quantum computing, are poised to exponentially increase the capabilities of generative models. We are on the brink of breakthroughs in encoding, simulation, and optimization that could catapult AI-generated content to levels of complexity and realism we've only dreamed about. Picture near-instantaneous rendering of hyper-realistic virtual worlds or the generation of novel pharmaceuticals with unprecedented precision. This isn't just science fiction—it's a potential future reality that comes closer with each technological stride.

Interdisciplinary synergy is another fertile ground for innovation. The merging of fields like neuroscience and AI can lead to models that not only mimic human creativity but also understand and replicate the nuances of human cognition. Imagine generative AI that "thinks" like a human, creating art and solving problems with the intuitive flair that characterizes human intelligence. Such advancements come with profound implications, from revolutionizing creative endeavors to tackling complex scientific challenges.

In the realm of virtual and augmented reality, generative AI will play a crucial role in creating more immersive experiences. The ability to generate real-time, adaptive environments means that users could explore endlessly unique landscapes, each dynamically crafted to their preferences and actions. This could revolutionize gaming, education, and even remote work, providing experiences that are tailored and updated continuously by sophisticated AI algorithms.

Generative AI is also likely to shift traditional notions of authorship and originality. Collaborative projects between humans and AI could become the new norm, challenging established ideas of what it means to be a creator. This emerging paradigm invites a rethinking of intellectual property laws and ethical considerations, becoming both a catalyst for new legal frameworks and a discussion point for the philosophical aspects of creation and ownership.

On a societal level, the accessibility of generative AI tools is likely to democratize creativity, opening up artistic and content creation possibilities to individuals who might not have had the opportunity before. Imagine a world where anyone can become an artist or a musician with the help of intuitive generative AI tools. This could lead to a flourishing of creativity, where diverse voices contribute to cultural and artistic landscapes enriched by a multitude of perspectives.

Another promising avenue lies in the medical field. Generative AI models can be employed to simulate complex biological systems and generate new hypotheses for treating diseases. For instance, generative design can assist in creating optimized prosthetics tailored to individual needs or in developing personalized treatment plans that evolve in response to a patient's progress. The applications extend beyond treatment to diagnostics, where AI-generated simulations can predict disease outbreaks or model the spread of infectious diseases under various scenarios.

Environmental sustainability could also receive a significant boost from generative AI. By modeling and optimizing renewable energy systems, AI-generated designs can lead to more efficient and sustainable energy solutions. Imagine urban planning aided by generative models that can simulate and optimize entire cities to minimize waste and energy consumption, leading to more sustainable living environments.

Looking at artistic domains, the fusion of AI with traditional art-forms promises to yield unprecedented hybrid creations. Generative AI can partner with artists not just as a tool but as a genuine collaborator, offering new techniques and inspirations. This synergy can birth entirely new genres of art that are not bound by human limitations, yet deeply enriched by human creativity. From music and literature to visual and performance arts, the future of creative expression stands to be radically transformed.

The educational landscape is another realm poised for transformation. Imagine generative AI systems that can create custom learning materials tailored to each student's learning style and pace. These AI tutors could offer personalized feedback and generate adaptive learning scenarios, making education both more effective and engaging. This could bridge gaps in education, offering high-quality, personalized learning experiences to students regardless of their socioeconomic status or geographical location.

Moreover, as AI continues to permeate various sectors, the demand for new skills and roles will emerge. Fields like AI ethics, AI-aided design, and AI system management will become crucial. Educational institutions may need to adapt, offering courses and programs that prepare students for careers in a world where generative AI is a staple. This could lead to a workforce adept in both creativity and technical prowess, ready to harness the full potential of generative AI.

Another fascinating direction involves the concept of AI-driven hyper-personalization of products and services. From custom-fit clothing designed through generative models to personalized travel itineraries that evolve based on real-time preferences and behaviors, the possibilities are limitless. Businesses could offer unprecedented levels of customization, creating products that not only meet but anticipate user needs in innovative ways.

In the sphere of entertainment, generative AI could lead to entirely new genres of films and video games. Think of interactive narratives where the plot evolves based on viewer preferences and choices, creating unique storylines for each user. This could redefine how we consume media, shifting from passive consumption to active participation.

Public infrastructure and urban development also stand to benefit from generative AI. Smarter cities with AI-generated traffic systems could alleviate congestion and improve quality of life. From generating waste management systems that optimize recycling to designing green spaces that promote community well-being, the application of AI in urban planning is bound to create more livable, efficient, and sustainable urban environments.

However, these advancements are not without challenges. The ethical considerations of AI-generated content, the potential for misuse, and the need for robust governance frameworks are issues that require serious attention. Responsible development and deployment of generative AI will be key to harnessing its full potential while mitigating risks.

In summary, the future of generative AI holds transformative potential across various domains. By seamlessly integrating into our tools and daily lives, enhancing creativity, solving complex scientific problems, and fostering sustainability, generative AI is set to redefine the boundaries of what is possible. While challenges remain, the promise of generative AI lies in its ability to innovate, inspire, and improve the human experience in ways that we are just beginning to imagine.

Chapter 22:
Troubleshooting Common Issues

As you dive deeper into generative AI, encountering obstacles is inevitable, but each challenge brings you closer to mastery. Troubleshooting involves a combination of methodical problem-solving and creative thinking to address issues like unexpected errors, model performance, and data discrepancies. Start by identifying the root of the problem—whether it's related to coding, data, or the model's architecture. Systematically test different components, check for common errors such as data formatting issues or algorithmic inconsistencies, and revisit your model's parameters and settings. Equally important is tapping into the community for support; forums and online groups can offer insights and solutions that you might not have considered. Ultimately, persistence and a proactive approach to learning from each setback will enhance both your skills and your projects.

Dealing with Errors

When delving into the captivating world of generative AI, you're bound to encounter some frustrations along the way. One moment your model is churning out awe-inspiring art, and the next, you're scratching your head over unexpected glitches. Understanding and addressing these errors is critical to both your growth and the quality of your output.

Errors can range from simple bugs in your code to more complex issues like model convergence problems. The first step to dealing with any error is identifying its nature. Is the issue related to the code, the data, or the model itself? Narrowing down the categories can save you a lot of time and effort.

Code Errors are often the easiest to fix but the trickiest to spot. Syntax errors, missing libraries, or incorrect configurations can all lead to your model failing to run as expected. Basic debugging tools and techniques such as print statements, debugging in your Integrated Development Environment (IDE), or even a simple code review, can help you catch these errors. An overlooked element such as a misplaced parenthesis or a typo in a variable name can make all the difference.

More vexing are **Data Errors**. The quality of the data you feed into your model plays a pivotal role in the outcome. Missing values, inconsistent data formats, or simply having too little data can significantly impact your model's performance. It's essential to rigorously clean and preprocess your data. Look for anomalies, remove duplicates, and ensure your data is properly labeled. Tools like pandas in Python can be invaluable for these tasks.

When dealing with *Data Errors*, consider augmenting your dataset with synthetic data if it's too small. Data augmentation techniques can include adding slight variations, like rotation or scaling to images, to generate more training examples. This approach can often enhance the model's ability to generalize better and mitigate overfitting.

On the more advanced end of the spectrum are **Model Errors**. These occur when something goes awry with how your generative model learns from the data. Model errors can manifest as poor convergence, where your model fails to improve over epochs, or mode collapse in GANs, where the generator produces the same output across different inputs. Tweaking hyperparameters such as learning rate, batch size, or trying different architectures, can sometimes solve these issues. Regularization techniques like dropout or weight decay can also help in improving the model's stability.

Chris Elliott

It's crucial to keep an eye on your loss functions and accuracy metrics. If you're observing a decreasing training loss but stagnant or increasing validation loss, you're likely witnessing overfitting. Conversely, if both training and validation losses aren't decreasing, your model might be too simplistic for the given task.

Errors aren't merely obstacles; they are learning opportunities. Take for instance a GAN that continuously generates blurry images. This isn't just a failure; it's a signal that something in the training process needs adjustment. Reflect on possible causes: is the discriminator too powerful compared to the generator? Perhaps tweaking the architecture or adopting techniques like one-sided label smoothing can alleviate the problem.

In some cases, errors might arise due to **Library and Environment Issues**. Dependency conflicts, outdated versions, or even operating system variances can lead to unexpected behavior. Using virtual environments or containers like Docker can help manage these dependencies more effectively, ensuring that your development environment is both consistent and reproducible.

Community support can be a goldmine when dealing with errors. Whether through forums, social media groups, or even academic papers, someone else has likely encountered and solved a similar issue. Online platforms like Stack Overflow, GitHub, or dedicated AI and machine learning communities can offer invaluable insights. Don't be afraid to ask questions and share your experiences; the collaborative nature of these communities often fosters better solutions.

Documenting your error-handling strategies is equally important. Keeping a log of problems encountered and solutions found can serve as a helpful reference for future projects. This practice not only saves time but also empowers you to tackle increasingly complex challenges with confidence.

168

Lastly, remember that dealing with errors is part of the creative process in generative AI. It's a journey from frustration to discovery. Each resolved error brings you one step closer to refining your model and achieving mesmerizing results. Embrace these challenges as opportunities to deepen your understanding and to innovate further. With persistence and a methodical approach, you'll find that overcoming errors not only enhances your technical skills but also adds to the richness of your creative endeavors.

Best Practices

When venturing into the realm of generative AI, it will undoubtedly involve a period of trial and error. To transcend these inevitable challenges effectively and efficiently, adhering to certain best practices can make a monumental difference. From understanding common pitfalls to leveraging advanced debugging strategies, these practices can significantly increase the likelihood of success while reducing frustrations. So let's delve into what it means to approach troubleshooting with the right mindset and techniques.

First and foremost, *understanding the importance of documentation* is key. Keeping detailed logs of your work, including the datasets you used, hyperparameters for your models, and any transformations applied, can save countless hours later on. Documenting not only helps you retrace your steps but also assists in identifying where things might have gone wrong. This practice becomes extremely useful in collaborative settings where other team members might need to understand your process.

Another cornerstone of effective troubleshooting is **systematic experimentation and iteration**. Rather than making multiple changes at once, adjust one parameter at a time and observe the outcomes. This controlled approach makes it easier to pinpoint the sources of issues. For instance, if the quality of generated images from a

GAN drops after a specific change, backtracking becomes much simpler when you have a clear record of single-parameter adjustments.

In parallel, it's crucial to *utilize robust evaluation metrics*. Evaluate your models using multiple metrics to gain a comprehensive understanding of their performance. For example, while a visual inspection of generated images can provide some insight, leveraging metrics like Fréchet Inception Distance (FID) or Precision and Recall can offer quantitative support to your subjective evaluations. In text generation, BLEU scores or perplexity might be more appropriate metrics to gauge quality.

Don't ignore the value of *community input and peer review*. Online forums, dedicated to AI and machine learning, such as GitHub, Stack Overflow, or specialized subreddits, can be invaluable resources for troubleshooting. Engaging with these communities not only helps in solving immediate problems but also fosters learning through shared experiences and advice.

Moving forward, make use of **comprehensive visualization tools**. Tools like TensorBoard, Matplotlib, and seaborn can help visualize loss curves, accuracy metrics, and other critical aspects of model performance. Visualization aids in quickly identifying anomalies such as vanishing or exploding gradients, which might not be obvious through raw numbers alone.

Don't underestimate the power of **documented best practices** specific to the tools and libraries you use. Libraries like TensorFlow, PyTorch, and others often come with extensive documentation that includes common pitfalls and recommended practices. Taking the time to read through these can preempt a lot of issues. Moreover, many of these platforms offer community support and forums where frequent issues and their solutions are discussed in detail.

A significant part of troubleshooting involves *effective data management.* Ensuring that your data is clean and correctly labeled is paramount. Inconsistencies in your dataset can lead to misleading model outputs and harder-to-diagnose problems. Employ techniques like data augmentation, normalization, and validation splits to maintain high-quality data. Automation tools for data pipelines can also streamline this process, making it more reliable and less prone to human error.

Consider implementing **version control systems** for your models and datasets. Tools like Git can keep track of different iterations of your models, making it easier to roll back to a previous state if a new change doesn't work as expected. Version control systems help in maintaining coherence and consistency across various stages of your project, especially when dealing with team collaborations and long-term projects.

In addition, the practice of **incremental training** should not be overlooked. Rather than training your models from scratch each time, consider incremental learning approaches where the model is continuously trained and fine-tuned over previous iterations. This method not only speeds up training times but also leverages previously learned features, thus enhancing model performance and stability.

Moreover, embrace the practice of *regularly updating your libraries and frameworks.* Machine learning is a rapidly evolving field, and keeping your tools up-to-date can help you leverage the latest features and optimizations. However, always test your models thoroughly when migrating to new versions to ensure compatibility and consistent performance.

Lastly, cultivating a mindset of **continuous learning and adaptation** is essential. The field of generative AI is evolving at a breakneck pace, with new research papers, techniques, and tools being developed constantly. Staying updated with the latest advancements through academic journals, conferences, and online courses can provide you with fresh perspectives and innovative solutions to common problems.

To summarize, effective troubleshooting in generative AI involves a blend of meticulous documentation, systematic experimentation, robust evaluation, community collaboration, comprehensive visualization, and continuous learning. By adhering to these best practices, you can navigate the complex landscape of generative AI with more confidence and less frustration, paving the way for more innovative and impactful projects.

CHAPTER 23:
CUSTOMIZING MODELS FOR PERSONALIZED ART

Customizing models for personalized art is about tailoring AI algorithms to reflect your unique vision and style, transforming a generic tool into a specialized assistant that resonates with your artistic voice. It goes beyond merely adjusting a few parameters; it's about understanding the intricacies of the chosen model, from neural network structures to training datasets, and making calculated modifications that can drastically alter the output. By fine-tuning parameters like learning rate, hyperparameters, and layers, artists can guide the AI to emphasize certain colors, patterns, or textures that are central to their creative identity. Moreover, personalization techniques such as transfer learning, where a pre-trained model is refined with your own dataset, enable the infusion of personal aesthetics and themes. This chapter is a bridge to unlocking an entirely new realm of creative expression, where the interplay between human intuition and machine calculation results in artwork that is both technologically sophisticated and deeply personal.

Tuning Parameters

When customizing generative AI models for personalized art, tuning parameters is a crucial step that can significantly impact the model's performance and the uniqueness of the produced artwork. Tuning parameters, also known as hyperparameters, involves adjusting various settings within the model to optimize its behavior according to your artistic vision. This process can be both an art and a science, requiring a balance between empirical experimentation and intuitive creativity.

The process begins by understanding the basic elements you have control over. These include learning rate, batch size, epochs, and other architecture-specific variables like layers and nodes in neural networks. The learning rate determines how quickly the model learns from the data during training. A higher learning rate can lead to faster training but risks overshooting the optimal solution, while a lower rate ensures more precise adjustments at the cost of extended training time.

Similarly, batch size refers to the number of data samples the model trains on before updating its internal parameters. Choosing a large batch size can speed up training but may lead to less generalizable results. On the other hand, a smaller batch size can capture subtler patterns in the data, enhancing the model's ability to produce more detailed and intricate art pieces. Fine-tuning these parameters requires careful consideration and iterative testing.

Another essential parameter is the number of epochs—basically, how many times the model sees the entire dataset during training. More epochs allow the model to learn better and refine its ability to generate art but run the risk of overfitting, where the model performs well on the training data but poorly on new, unseen data. To counter overfitting, techniques like dropout (randomly omitting certain nodes during training) and L2 regularization (adding a penalty for large weights) play a vital role in this balancing act.

Different models also offer specialized parameters. For instance, when using GANs (Generative Adversarial Networks), one must carefully tune the balance between the generator and discriminator components. This involves adjusting the rate at which these two networks compete against each other during training. If the discriminator becomes too strong, the generator struggles to produce convincing samples. Conversely, a weak discriminator cannot adequately challenge the generator, leading to subpar creations.

In VAEs (Variational Autoencoders), the tuning process includes setting the right balance between the reconstruction loss and the KL-divergence term. This balance significantly influences how well the model can create variations of the input data. If too much emphasis is placed on reconstruction loss, the model may simply memorize the training data, while a focus on KL-divergence aids in generating diverse and novel outputs.

For autoregressive models and flow-based models, hyperparameters like the order of autoregression and the type of coupling layers, respectively, become focal points. Adjusting these intricacies determines the quality and creativity of the resulting artwork. All these nuances collectively highlight that tuning parameters is not a mere technical task but an exercise that requires one to think like both a scientist and an artist.

Exploratory data analysis (EDA) is invaluable here. Before diving into parameter tuning, it's beneficial to conduct EDA to grasp the characteristics of your data—its distribution, key patterns, anomalies, etc. Visualizing the data can give you insights that inform better parameter choices. For example, if your dataset of abstract paintings shows a wide range of color palettes, you may choose a model architecture and tuning strategy that maximizes the diversity in the output.

In practice, tuning parameters often involves a method known as grid search or randomized search. Grid search systematically explores a predefined set of hyperparameters, testing each combination to find the optimal setting. While comprehensive, this approach can be computationally intensive. Randomized search offers a more efficient alternative by randomly sampling from the hyperparameter space, which often yields similarly good results with reduced computation.

Advanced techniques like Bayesian optimization provide an even more sophisticated approach to hyperparameter tuning. By building a probabilistic model of the function that maps parameter settings to their performance, Bayesian methods can predictively explore the most promising areas of the parameter space. This approach not only speeds up the tuning process but also enhances the model's final performance.

Furthermore, utilizing tools such as TensorBoard or other visualization software can be immensely helpful in monitoring the tuning process. These tools offer real-time insights into how changes in hyperparameters affect the model's training dynamics, allowing for more informed adjustments. They also help in spotting issues like vanishing gradients or overfitting early in the process, making corrective actions more straightforward.

It's also essential to approach parameter tuning iteratively. Start with broader sweeps through the parameter space to identify rough ranges that work best, followed by finer adjustments. Patience and persistence are key here, as finding the right combination often takes multiple rounds of trial and error. Documenting each iteration, the parameters tested, and the results obtained can significantly streamline this process, providing a clear path to follow in subsequent tuning efforts.

Finally, as you gain experience, you may develop an intuition for what settings work well for particular types of generative tasks. However, it's crucial to stay open to new techniques and emerging best practices in the field. Generative AI is a rapidly evolving discipline, and continuing education through forums, research papers, and collaborative projects can keep your skills at the cutting edge.

In summary, tuning parameters is a delicate balancing act that combines both technical proficiency and artistic sensibility. By meticulously adjusting hyperparameters, leveraging advanced tuning techniques, and remaining adaptive, you can significantly enhance the creativity and quality of your personalized AI art models. This process not only enriches your understanding of the underlying technology but also opens new avenues for creative expression and innovation.

Personalization Techniques

Personalizing generative AI models to create custom art offers an exciting field of exploration, where technology meets individual creativity. Whether you're a beginner or an AI art enthusiast, personalization allows you to weave your unique touch into the art pieces generated by algorithms. The magic lies in making the output look less generic and more reflective of a specific vision, style, or emotion.

To start with, one of the primary steps in personalization is data selection and preparation. The data you feed into your model can significantly influence the outcome. For instance, if your artistic preference leans towards impressionism, you can train your model on a dataset rich with impressionist paintings. By carefully curating your dataset, you're setting the foundation for your model to understand and replicate those artistic nuances.

Next up, tweaking the model architecture itself can lead to more personalized results. You might want to experiment with different types of neural networks or layer compositions to see how they impact the final output. Fine-tuning, a subset of transfer learning, involves starting with a pre-trained model and adjusting only certain layers or parameters. This technique is not only efficient but also leverages existing knowledge encoded in the pre-trained model, providing you a head start in generating personalized art.

Parameter tuning is another powerful tool. By adjusting parameters such as learning rate, epoch numbers, and batch sizes, you can influence how your model learns and performs. For example, a lower learning rate might slow down the training process but can lead to more detailed and fine-tuned results. Conversely, a higher learning rate can expedite training but might result in less intricate outputs. Finding the right balance is often an iterative process requiring both patience and experimentation.

Control mechanisms, such as style transfer, can also significantly contribute to personalization. Style transfer involves training a model to apply the style of one image to the content of another. This technique can be utilized to merge your favorite artworks' styles with new content, creating entirely original pieces that still bear your distinct artistic signature. It's like having a digital brush that paints in your style every time.

A promising method for personalized art creation is using conditional generative models. Conditional Generative Adversarial Networks (cGANs) allow you to condition the output on auxiliary information. This means you can guide the model to generate art based on specific features or criteria, like color palette, thematic elements, or even more abstract qualities such as mood. This makes the process interactive and offers you more control over the final result.

Interactivity is another cornerstone of personalization. You can use interactive user interfaces to make on-the-fly adjustments as you observe the generative process. These interfaces allow for real-time parameter tuning, enabling an iterative and exploratory approach to creating art. Many modern tools come with user-friendly interfaces where sliders and buttons let you tweak aspects ranging from color saturation to pattern complexity.

Data augmentation techniques can also be leveraged to add a personal touch. By augmenting your dataset with transformations such as rotations, translations, and scaling, you can enrich the learning experience of your model. This not only makes the model more robust but also introduces subtle, yet significant variations that help in generating unique art pieces. Essentially, you're teaching the model to understand and replicate a broader range of possibilities.

Another exciting avenue for personalization is using human feed-back loops. Interactive approaches allow users to rate or select pre-ferred outputs, which are then fed back into the system for further re-finement. This closed-loop system can significantly enhance the quali-ty and relevance of the generated art, making it more aligned with the user's tastes and preferences.

Then there's the aspect of employing custom loss functions. Tra-ditional generative models use predefined loss functions to guide the learning process. However, you can define custom loss functions tai-lored to your artistic goals. For instance, if you value certain stylistic elements more than others, your loss function can be adjusted to pe-nalize deviations from those elements more heavily. This ensures that the generated art adheres more closely to your envisioned style.

Generative models also offer the flexibility of experimentations with latent space manipulations. By exploring the latent space—the abstract multi-dimensional space where your data resides—you can steer the model towards generating specific types of outputs. Tech-niques such as interpolation between points in the latent space can produce a continuum of outputs, allowing you to explore a spectrum of styles or themes. This method brings a level of depth and personali-zation that's akin to discovering hidden gems within your own creative potential.

Custom training routines sometimes come into play for achieving specific artistic goals. This could involve implementing alternative training schedules, freezing certain layers while training others, or employing novel optimization techniques. These tailored routines can result in highly personalized models that often outperform generic training regimes in terms of capturing individualistic styles and nuances.

Personalization doesn't stop at visuals; it also extends to perfor-mance and efficiency adjustments. Models can be optimized to work on specific hardware setups, making them more accessible for real-time applications. Techniques like model pruning or quantization can help streamline computational requirements, ensuring that even personalized models run efficiently on less powerful machines. This is particularly useful when integrating AI art into interactive installations or real-time applications.

Incorporating multimodal inputs, like combining text, images, and even sound, can add another layer of personalization. For example, guiding a visual art-generating model with descriptive text inputs can help in creating pieces that fit a narrative or thematic direction you're interested in. This cross-modal approach broadens the scope of what can be achieved, allowing your artistic expression to transcend traditional boundaries.

Collaborations between humans and machines often further enrich the personalization process. Providing creative prompts or partially complete inputs for the generative model to finish can lead to hybrid art forms that blend human creativity with machine precision. These collaborative efforts can result in unique and unrepeatable works of art that highlight the strengths of both human intuition and AI capability.

The emotional aspect of art should not be overlooked in the personalization process. Tools that allow you to tag images with emotional labels or categories can guide the generative process to produce art that resonates on a deeper level. By aligning generated art with specific emotional tones, you're able to create pieces that not only look great but also evoke intended feelings or memories.

Finally, ongoing adaptation is crucial for maintaining the relevance of personalized models. As your artistic interests evolve, you can continually update and refine your model to keep it aligned with your current tastes. This could involve periodic retraining with new datasets, incorporating new styles, or adjusting parameters to reflect your latest inspirations. This dynamic process ensures that your personalized art remains fresh and continuously resonates with your evolving creative vision.

Personalizing generative models for art is a rich, multifaceted journey. It combines technical know-how with artistic intuition, offering a playground where your imagination can truly run wild. With an array of techniques at your disposal, the possibilities for making generative AI art that's uniquely yours are virtually endless. Dive in, explore, and let your creativity soar.

CHAPTER 24:
COLLABORATIONS BETWEEN HUMANS AND MACHINES

The synergy between human creativity and machine intelligence opens up a realm of hybrid art forms that defy traditional boundaries. By combining the nuanced touch of human artists with the computational power of AI, we can produce works that are both technically impressive and emotionally resonant. These collaborations enable artists to push the limits of their imagination, incorporating elements that would be impossible to achieve alone. From interactive installations that respond to audience input to generative music that evolves in real-time, the fusion of human and machine creativity paves the way for innovative, dynamic art experiences. The collaboration not only creates unique art but also inspires new methods and philosophies in artistic expression, raising intriguing questions about authorship and the essence of creativity.

Hybrid Art Forms

Hybrid art forms represent a fascinating intersection where human creativity meets the computational prowess of machines. This meeting ground has resulted in some of the most innovative and thought-provoking pieces of art in recent years. The collaboration between humans and machines has not merely augmented traditional art; it has birthed entirely new genres. These hybrid art forms are characterized by the harmonious integration of human intuition with machine intelligence, producing works that challenge our understanding of both art and technology.

The allure of hybrid art lies in its ability to transcend the limitations imposed by either creator alone. Artists no longer work in isolation; they co-create with algorithms capable of generating endless variations and iterations. While a human artist's sensibilities drive the conceptual aspects, the machine's algorithmic efficiency handles large-scale computations and pattern recognitions. This synergy results in art that is both deeply personal and extraordinarily complex.

For instance, consider the evolving field of generative visual art. Here, artists provide initial input, such as a dataset of images or specific stylistic parameters, and machines, often utilizing Generative Adversarial Networks (GANs) or Variational Autoencoders (VAEs), generate new images that push the boundaries of conventional aesthetics. What emerges are works that possess both the imprint of their human creators and the algorithmic "fingerprint" of the machine. These new forms often feature intricate complexities that would be difficult, if not impossible, for a human to devise unaided.

Music is another domain that has been significantly transformed by these collaborations. Through the use of AI models like OpenAI's MuseNet or Google's Magenta, musicians now have the ability to co-compose with algorithms that can understand and mimic various musical styles and genres. These AI models can create harmonies, suggest chord progressions, or even improvise solos, providing musicians with a virtual collaborator that expands their creative reach. The resulting compositions are a fusion of human emotional depth and machine-driven innovation.

Beyond traditional art and music, the realm of hybrid art forms has expanded into interactive and immersive installations. In such projects, AI plays a crucial role in creating responsive environments that adapt to the behavior and emotions of participants. Artists collaborate with teams of engineers and data scientists to design algorithms that interpret real-time data, including movement, speech, or physiological cues, and then react accordingly. These installations break the barrier between spectators and art, turning passive observers into active participants.

Further exploring this concept, hybrid digital sculptures represent another fascinating frontier. Here, artists leverage AI to transform digital data into tangible, three-dimensional forms. Techniques such as 3D printing combined with AI-driven design algorithms allow the creation of complex structures that can evolve and change over time. These sculptures often incorporate sensory feedback mechanisms, making them not only visually captivating but also interactive. The line between digital and physical art becomes increasingly blurred, providing a multi-dimensional experience.

Hybrid literature is yet another exciting development. Writers can now collaborate with AI to generate text that ranges from individual sentences to entire chapters, incorporating AI's ability to parse vast amounts of literary data and reproduce different writing styles. Tools such as AI Dungeon or GPT-3 have opened new possibilities for interactive storytelling, allowing readers to interact with narrations that adapt in real-time based on their inputs. The result is a dynamic, ever-evolving narrative that offers a personalized experience for each reader.

Furthermore, hybrid performance art combines the best of both worlds: human expressiveness and machine precision. Dancers can interact with AI systems that modify lighting, sound, and projections based on their movements, creating a deeply immersive experience. These performances often involve wearable technology that tracks physiological data such as heart rate and muscle tension, feeding this information into an AI system that adapts the performance environment in real time. This form of art emphasizes the unity between human and machine, producing a performance that feels both organic and technologically advanced.

Not to be overlooked, hybrid cinematic arts are also gaining traction. Filmmakers are using AI to edit footage, generate special effects, and even write scripts. For example, AI-driven software can analyze raw footage to recommend the best sequences for constructing a narrative. Additionally, AI can generate visual effects that are seamlessly integrated into live-action footage, creating filmic experiences that are both realistic and fantastical. This collaboration allows filmmakers to push creative boundaries while maintaining high levels of efficiency and precision.

It's worth noting that these hybrid forms aren't simply about combining human and machine efforts; they invite a dialogue between the two. This dialogue can raise essential questions about authorship, originality, and the very nature of creativity. When an AI generates a piece of art, is it the machine that should be credited, or the human who designed the algorithm? Or perhaps both? These questions challenge our traditional notions of what it means to create and open up philosophical debates that draw from both art theory and ethics.

The educational sector is also exploring hybrid art forms as tools for teaching and learning. Courses that combine elements of coding, design, and art encourage students to think across disciplines and to develop a hybrid mindset. By integrating AI into art education, institutions can prepare future artists to use these technologies as extensions of their own creativity, rather than viewing them as mere tools. This integrative approach not only broadens the scope of what students can achieve but also equips them to navigate the evolving landscape of art and technology.

The business world has not been left behind either. Companies are increasingly commissioning hybrid art projects to enhance brand engagement and customer experience. From AI-generated logos to interactive campaigns, businesses recognize the value of integrating artistic and technological innovations. These hybrid projects not only captivate audiences but also serve as powerful marketing tools that highlight a brand's modernity and forward-thinking approach.

Of course, hybrid art forms also come with their challenges. Issues such as the ethical use of data, the potential for algorithmic bias, and the risk of over-reliance on technology are all important considerations. Artists and developers must navigate these challenges thoughtfully, ensuring that their work adheres to ethical standards and promotes inclusivity. As hybrid art evolves, so too must our frameworks for understanding and evaluating its impact.

The future of hybrid art forms is both exciting and unpredictable. As advancements in AI continue, the possibilities for new forms of expression and creativity are boundless. We can expect to see an increasingly sophisticated interplay between human and machine contributions, leading to art that is more interactive, personalized, and immersive than ever before. This ongoing evolution promises to redefine our perceptions of art and creativity, challenging us to reconsider the boundaries between man and machine.

In conclusion, hybrid art forms represent an extraordinary confluence of human imagination and technological innovation. As artists and machines continue to collaborate, we'll undoubtedly see new genres and modalities emerge, each pushing the envelope of what art can be. This dynamic interaction serves as a testament to the limitless potential of creativity, inviting us to envision a future where the synthesis of human and machine capabilities redefine the very essence of artistic expression.

Case Studies

The creative landscape is continuously evolving, and the collaboration between humans and machines stands as one of the most exciting arenas in this cross-disciplinary journey. Let's delve into some pivotal case studies that highlight how generative AI and human creativity have converged to produce groundbreaking results. Each of these examples underscores not only the ingenuity of the technology but also the irreplaceable flair of human artists and developers.

One noteworthy case is the creation of the "Next Rembrandt" project. Spearheaded by a team comprising data scientists, developers, and art historians, this project aimed to generate a new artwork in the style of the famous Dutch painter, Rembrandt van Rijn. By analyzing high-resolution scans of Rembrandt's works, the team compiled a deep dataset that included the artist's brushstrokes, color palette, and subject matter preferences. Using this information, they trained a deep learning model to recreate a new painting that could seamlessly blend into Rembrandt's repertoire. The final output was a portrait of a man in traditional 17th-century attire, astonishingly rendered to the minutest detail. This project serves as a testament to how AI can augment traditional art forms, paying homage to historical styles while leveraging contemporary technology.

Another fascinating collaboration involved the Google Deep-Dream project. This initiative began as an internal experiment within Google to understand what neural networks perceive in images. By exaggerating the patterns recognized by the networks, the team generated surreal, dream-like images that turned the ordinary into the extraordinary. Artists quickly realized the potential of this tool for producing unique, psychedelic visuals. DeepDream has since been adopted by numerous artists who integrate its hallucinogenic aesthetic into various forms of media, from visual art to video installations. The project's success demonstrates AI's capability to inspire new artistic directions, transforming both the creation process and the resulting artwork.

In the music domain, the collaboration between human musicians and AI has also yielded fruitful results. For instance, the album "I AM AI" by Taryn Southern broke ground as the first entirely AI-composed album in history. Taryn collaborated with the AI platform Amper Music, which allows users to generate musical compositions based on their inputs regarding mood, length, and instrumentation. While Amper took care of the initial compositional groundwork, Taryn added lyrics, vocals, and final production touches, creating an emotionally resonant album that melds AI-generated melodies with human story-telling. This symbiotic relationship highlights the transformative pos-sibilities of AI in music, pushing the boundaries of what can be achieved when human intuition and machine precision come together.

Film has also seen significant innovation through AI collaboration. Take, for example, the script for "Sunspring," a short science fiction film entirely written by an AI named Benjamin, created by filmmaker Oscar Sharp and AI researcher Ross Goodwin. Benjamin was trained on hundreds of sci-fi scripts and, despite producing somewhat abstract and nonlinear text, it provided a unique narrative foundation for the film. The human filmmakers then interpreted and visualized this un-conventional script, resulting in a quirky yet thought-provoking film. This case illustrates the imaginative potential of AI in narrative art forms, challenging traditional storytelling paradigms and encouraging more experimental approaches.

In fashion, the collaboration between AI and designers has birthed collections that push aesthetic boundaries. One example is the work of German designer Hanna Smiatek, who utilized generative adversarial networks (GANs) to come up with new and innovative clothing designs. By training her models on a database of existing fashion items, she could generate entirely new pieces that combine elements from various styles and eras. These AI-curated designs presented possibilities that would be difficult to conceive manually, thus offering a fresh perspective on future fashion trends. The interplay between human creativity and machine learning in the fashion industry highlights how seemingly disparate fields can intersect to produce unprecedented innovations.

Another significant case study is the AI-driven project called "The Grid", a web design platform that uses machine learning to autonomously build websites based on user content and preferences. By analyzing design principles and user data, The Grid's AI can automatically choose layouts, color schemes, and typography that aesthetically align with the user's brand. This enables users to generate professional-grade websites with minimal input, democratizing access to high-quality web design. While designers initially viewed such technologies as competition, many quickly realized their potential as powerful tools that could streamline their workflow and broaden their creative capabilities.

The collaborative use of AI in interactive installations provides another compelling case study. For instance, teamLab, a Tokyo-based interdisciplinary art collective, employs AI to create immersive digital environments. Their installations, such as "teamLab Borderless," utilize AI to produce dynamic artworks that change in real-time based on viewer interaction. As visitors move through the exhibit, the AI alters patterns, colors, and sounds accordingly, creating a deeply personalized and ever-evolving art experience. This marriage of human creativity and machine learning invites audiences to participate actively in the artistic process, making the art itself a living, breathing entity shaped by collective human interaction.

Perhaps one of the more ethically nuanced case studies is the "AI Gaydar" project created by Yilun Wang and Michal Kosinski at Stanford University. They trained a neural network to predict sexual orientation based on facial images, sparking a heated ethical debate about the scope and implications of AI. While technically sophisticated, the project raised serious questions about privacy, consent, and the potential for misuse. This case is a stark reminder that collaboration between humans and machines requires vigilant ethical considerations, ensuring that technological advancements serve to benefit society rather than harm it.

Generative AI's role in journalism is another sphere worth mentioning. Automated Insights, a company specializing in natural language generation, introduced their platform called Wordsmith to automate the production of news stories, particularly in data-heavy contexts like financial reporting and sports journalism. Rather than replacing journalists, this technology acts as an accelerant, allowing them to focus on investigative and narrative-driven work while the AI handles routine reporting. This balance of labor exemplifies a productive collaboration where both humans and machines bring their unique strengths to the table.

In the realm of architecture, firms like Zaha Hadid Architects have begun integrating AI into their design processes. By using generative design algorithms, architects can explore hundreds of design iterations quickly, factoring in environmental considerations, material constraints, and aesthetic guidelines. This allows for a more holistic and efficient approach to building design. The symbiosis of human architectural expertise and AI-driven optimization leads to structures that are not only visually stunning but are also environmentally sustainable and economically viable.

In healthcare, the collaboration between humans and AI has led to advancements in diagnostic imaging. For instance, Google's Deep-Mind has developed AI systems that can analyze retinal scans to detect early signs of diseases like diabetic retinopathy and age-related macular degeneration with high accuracy. While these systems serve as invaluable tools, the role of the human physician remains irreplaceable for final diagnosis and treatment planning. This partnership underscores how AI can enhance medical care by providing precise, data-driven insights that support clinical decision-making.

As we've seen, whether it's through generating art, composing music, designing fashion, creating films, formulating web designs, or even pushing ethical boundaries, the collaboration between humans and machines has led to extraordinary innovations

CHAPTER 25:
LEGAL ASPECTS OF GENERATIVE AI

The legal landscape surrounding generative AI is a complex and evolving terrain, with significant implications regarding copyright, fair use, and intellectual property. As beginners and enthusiasts explore the creative potential of generative AI, it is crucial to understand how legal frameworks apply to AI-generated works. These considerations are not merely theoretical; they have real-world impact on how creators and users interact with AI technologies. For instance, questions arise about who owns the rights to artwork generated by AI— is it the programmer who designed the algorithm, the user who inputs data, or the AI itself? Additionally, existing copyright laws and their interpretation in the context of AI-generated content can vary, making it essential to stay informed about recent legal cases and precedents. Navigating these legal waters requires a blend of vigilance and adaptability, ensuring that creative endeavors in the generative AI space are both innovative and compliant with prevailing laws and regulations. By diving into the legal aspects, we aim to equip you with the knowledge to protect your work and respect others' rights, fostering a responsible and ethical approach to using generative AI in your creative projects.

Copyright and Fair Use

The intersection of copyright law and generative AI art is a complex and evolving landscape. At its core, copyright is a legal framework designed to protect the original creations of authors, artists, musicians, and other creators. When it comes to generative AI, the question of who holds the copyright—the programmer, the user, or even the AI itself—becomes increasingly complicated. Understanding these nuances is essential for any beginner or enthusiast interested in diving into the world of generative art.

Traditionally, copyright law grants the creator of an original work specific exclusive rights, such as the right to reproduce, distribute, and display the work. But who is the creator when an AI program generates the artwork? In most jurisdictions, copyright law doesn't yet recognize AI as holding the capacity for legal authorship. As a result, the humans behind the AI—either the developers who created the algorithms or the users who input data and parameters—are usually considered the de facto creators and copyright holders. However, this interpretation can vary significantly depending on the legal framework of the country in question.

Another critical aspect to consider is the training data used to develop generative AI models. These datasets often comprise millions of copyrighted images, texts, or music tracks. Using this data to train an AI model can raise significant copyright issues, especially if the output generated by the AI closely resembles its training examples. If an AI-generated piece looks too similar to existing copyrighted works, it may infringe on those original works' copyright, even if the resemblance was unintentional.

Fair use is another concept that can come into play when discussing generative AI. Fair use permits limited use of copyrighted material without acquiring permission from the rights holders, provided that the use meets specific criteria. In the context of generative AI, this could include using copyrighted images or text for research, education, or parody. However, defining what constitutes fair use can be highly subjective and varies depending on the jurisdiction. For instance, an AI art piece used for educational purposes in the United States might be classified as fair use, whereas, in other countries, it might not receive the same protection.

Transformative use is a central pillar of the fair use doctrine. If an AI-generated work transforms the original input in a way that adds new expression, meaning, or message, it may be considered transformative and thus fall under fair use. For instance, using a generative AI to create entirely new images that merely reference the style or elements of the original images could be considered transformative. However, this is a legal grey area, and litigation is often required to determine whether a specific use qualifies as transformative fair use.

Several landmark legal cases have already begun to address some of these concerns. One notable case involves the use of AI to generate art from existing copyrighted works. The courts are grappling with questions around whether the new art is sufficiently transformative to merit its own copyright protection or if it merely constitutes a derivative work that still infringes on the original creator's rights. Although each case has its unique circumstances, the outcomes of these legal battles will set precedents for how copyright law is interpreted in the context of generative AI.

The role of contracts and licenses is also paramount in the realm of generative AI. When using third-party datasets or pre-trained models, it is essential to understand the terms and conditions laid out by the original creators or distributors. These legal agreements often specify how the data or models can be used, modified, and shared. Ignoring these terms can lead to legal repercussions and infringe on the rights of the original copyright holders.

As a generative AI enthusiast or beginner, it may be prudent to consult a legal expert when exploring new projects, especially if you intend to commercialize your creations. Legal advice can help you navigate the complexities of copyright and fair use, ensuring that you stay within the legal bounds and protect yourself from potential lawsuits. Additionally, understanding these legal frameworks will empower you to make informed decisions about your creative processes, from selecting training data to sharing or selling your AI-generated art.

Another route to navigate these complexities is to use open-source datasets and models. Open-source platforms often come with licenses that are more permissive, allowing for greater freedom in experimentation and creation. By choosing resources that are labeled for reuse and modification, you can lower the risk of infringing on copyright laws. However, even open-source materials come with their own sets of rules and restrictions that must be carefully reviewed.

Collaborative projects can also benefit from a clear understanding of copyright and fair use principles. When multiple parties contribute to a generative AI project, establishing clear agreements upfront regarding ownership and rights can prevent disputes down the line. These agreements should outline who holds the copyright, how any revenue will be shared, and how each party can use the final outputs.

In some sectors, voluntary registries and digital rights management (DRM) systems are emerging as tools to help manage and protect AI-generated works. These systems allow creators to register their works, establish proof of authorship, and control how their creations are used and distributed. While not legally binding, such registries can offer an additional layer of protection and help to establish a creator's rights in the digital age.

As the generative AI field continues to grow and evolve, so too will the landscape of copyright and fair use. Staying informed about the latest developments in law and technology is crucial for anyone interested in this space. Industry standards, ethical guidelines, and evolving legal precedents will all play a part in shaping the future of generative AI and its creative applications.

In conclusion, copyright and fair use are pivotal considerations for anyone delving into generative AI. Understanding these legal concepts will not only help you protect your work but also ensure that you respect the rights of other creators. As you explore the thrilling possibilities of generative AI, keeping these legal aspects in mind will allow you to navigate the creative landscape with greater confidence and integrity.

Legal Case Studies

Legal case studies are crucial for understanding how the courts interpret the complexities surrounding generative AI. Over the past few years, several key cases have shaped the legal landscape regarding AI-created works, setting precedents that future cases will likely follow. These cases help us see not just the letter of the law, but its real-world application and implications.

One of the landmark cases in the realm of generative AI is *Naruto v. Slater*. While the case did not directly involve AI, it revolved around the concept of copyright ownership for works created by non-humans. In this case, a monkey named Naruto took a photograph using a camera set up by nature photographer David Slater. PETA filed a lawsuit claiming that Naruto should own the copyright to the photograph. The court ultimately ruled that animals cannot hold copyrights. This case is often cited in discussions about whether AI, as non-human entities, can own the rights to their creations.

Another significant case is that of *Feist Publications, Inc., v. Rural Telephone Service Co.* Even though it dates back to 1991, this case set the precedent for copyright law regarding originality. The Supreme Court ruled that raw data or factual information not governed by creativity is not copyrightable. This ruling has implications for AI, as many generative AI systems produce work by compiling existing data.

Understanding the fine line where creativity begins and raw data ends is crucial for cases involving AI-generated content.

In a more recent and directly relevant instance, the case of *Thaler v. the United States Copyright Office* tackled the issue head-on. Stephen Thaler applied for copyright on behalf of an AI system named "Creativity Machine" for an artwork the AI had created. The U.S. Copyright Office rejected the application, citing that only works created by human beings qualify for copyright protection. This decision underscores a critical roadblock for those looking to commercialize AI-generated art: the human authorship requirement.

Maryland v. King and its aftermath further explore the implications of AI in legal frameworks. While the case centered around DNA testing and forensic science, its broad acceptance of technological advancements in evidence gathering has opened doors for AI-generated data to be considered in court. It suggests that while AI can't own copyrights, the outputs of AI, such as data, may still be variously protected under different aspects of the law.

One fascinating example is the European Union's case *Infopaq International A/S v. Danske Dagblades Forening*. This involved the automatic generation of text snippets by an AI system. The EU Court found that even text snippets produced by software could be subject to copyright if they met the required threshold of originality. This ruling indicates that AI-generated content might get some form of protection if it resembles enough human-produced work to meet originality standards.

Then there is the curious case of *Aleph Farms* and *Technion-Israel Institute of Technology*, where the legality of lab-grown meat was scrutinized. While not strictly about generative AI, this case shows how emerging technologies can challenge existing legal frameworks. It highlights the necessity for new laws and regulations to account for technological advancements, setting the stage for future discussions about AI in courtrooms.

The recent case *Getty Images v. Stability AI* raises another dimension of complexity. Getty Images accused Stability AI of scraping millions of images from its site to train its generative models without permission. This case touches on the ethical use of copyrighted material to train AI. If the court sides with Getty Images, it could significantly impact how datasets for generative AI are created and used.

Taking another turn, the case *Atari v. Redbubble* explored the boundaries of automated designs and trademark rights. Redbubble's platform allows users to create and sell products featuring AI- generated designs. Atari argued that some designs infringed on its trademarks. The court's decision to side with Atari indicates a willingness to uphold traditional IP laws, even in the face of complex AI-generated content.

A novel yet related case involved artist *Robin Sloan*, who created a novel using his own text-generation model. When Sloan attempted to publish the novel, questions arose about authorship and copyright. This led to discussions rather than a legal altercation, but it nudged the legal community to consider how to attribute authorship when human and machine efforts are intertwined. While no formal case came from this, it set the stage for future disputes.

Turning to a different legal system, Japan's copyright dialogue involves a unique case with Sony Computer Science Laboratories and their AI-generated music. Here, the issue isn't about authorship alone but also about how profits are shared and responsibilities are allocated. Given Japan's progressive take on technology, the resolution might offer alternative precedents, compared to Western jurisprudence.

The China-based case of *Tencent vs. ByteDance* put two tech giants in opposition over AI-generated literature. Tencent argued that

ByteDance's AI was generating stories too similar to those housed in its own directories. The ruling reflected China's evolving yet stringent copyright laws, which are increasingly being tested by burgeoning AI technologies. Here we see another example of how different legal systems handle similar AI challenges.

Lastly, it's important to mention instances where legal cases aren't directly held in courts but under regulatory bodies. The European Union's General Data Protection Regulation (GDPR) has been used to address complaints about data misuse by AI systems. For instance, several anonymous cases have cited GDPR violations regarding how generative AI uses personal data to create new content.

These cases underscore a fundamental challenge: Legal systems worldwide are grappling to keep pace with the rapid advancements of AI technology. Each case that comes up is a foray into uncharted legal territory. Interpretations of current laws vary, making it a buffet of precedents from which future cases can draw – supporting either side of any given argument.

To summarize, these legal battles emphasize the need for updated laws and a comprehensive understanding of how generative AI can fit within existing legal frameworks. They provide instructive lessons, not just in the rhetoric of the courtroom but also in the very fabric of how society values creativity and ownership in a digital age. They showcase different approaches - from strict adherence to existing laws to creative interpretations, and even to discussions about new legal paradigms. With each ruling and regulation, we move a step closer to a more nuanced, informed dialogue about generative AI and its place in our legal systems.

CONCLUSION

The journey through the realms of generative AI, from the foundational elements to the zenith of creative applications, is an odyssey of vast potential and endless exploration. As we've traversed the topics ranging from understanding the basics of machine learning to delving into the nuances of generative adversarial networks and variational autoencoders, it's clear that this domain is not only technically rich but also ripe for unprecedented creative opportunities.

Generative AI stands as a testament to the symbiosis between human ingenuity and machine capability. The collaboration yields creations that neither entity could achieve alone. This synthesis makes AI not just a tool, but a collaborative partner in artistic creation. From producing awe-inspiring visual arts to composing soulful music and crafting indicative text, the possibilities are as vast as the expanse of human imagination.

One of the most profound aspects of generative AI is its democratization of creativity. Now, more than ever, both budding artists and seasoned professionals can unlock new avenues for expression without being held back by traditional skill limitations. With the plethora of accessible tools and libraries, setting up a creative AI workspace has become feasible for anyone with a computer and an internet connection. The era of exclusive artistic production is shifting towards inclusivity and diversity, fueled by generative AI.

However, with great power comes great responsibility. Ethical considerations are paramount as we explore and exploit these new technologies. It is crucial to remain vigilant about data privacy, ethical sourcing, and the potential biases within AI systems. As artists and creators, we must ensure that generative AI not only enhances creativity but also respects and preserves the fundamental values of originality and authenticity.

The collaborative endeavors between humans and machines bring about a fascinating new genre of hybrid art forms. These forms challenge traditional boundaries and invite us to reconsider what art can be in the digital age. Case studies and real-world projects showcased throughout this book underscore the incredible impact that generative AI can have, pushing the envelope of what is technologically and artistically possible.

Engaging with the AI art community further fuels this emergent field. By connecting with like-minded individuals through online forums, groups, conferences, and events, one can exchange ideas, seek inspiration, and find support. This collaborative spirit not only enhances individual projects but advances the field as a whole.

The future of generative AI is both exhilarating and uncharted. Emerging technologies and innovations portend a future where AI's role in art is not just supplemental but integral. As we venture forward, the art created by AI will undoubtedly evolve, becoming more sophisticated and integrated into mainstream art forms. Predicting these trends allows us to stay ahead of the curve and continue innovating with intention and foresight.

Troubleshooting common issues, understanding the legal landscape, and customizing models will remain critical as you continue your explorations. Mastery of these areas will not only resolve roadblocks but will also empower you to push the technical and creative boundaries of your projects. Building and sustaining a framework for continuous improvement ensures that your generative art remains dynamic and transformative.

Chris Elliott

Monetizing AI art opens up intriguing possibilities for artists looking to commercialize their creations. Whether it's through licensing, sales, crowdfunding, or sponsorship, the avenues for financial reward are as diverse as the art forms themselves. Understanding these pathways can turn passion projects into sustainable artistic careers.

The potential for generative AI to revolutionize art is immense. This book has sought to equip you with the knowledge and tools necessary to dive deeply into this evolving field. Whether you are a beginner taking your first steps or an enthusiast keen to expand your toolkit, the fundamentals covered herein are designed to empower you to craft, iterate, and innovate.

As we conclude our exploration, remember that the world of generative AI is ever-changing and constantly evolving. Stay curious, stay passionate, and most importantly, keep creating. The true beauty of generative AI lies not just in the technology itself, but in what you, as artists, can imagine and bring to life. The horizon is as broad as your creativity allows—go and make something wonderful.

Appendix A:
Appendix

In this appendix, we've compiled supplemental information and re-sources to enhance your journey into the fascinating world of genera-tive AI. While the primary chapters provided in-depth discussions and thorough explorations of various aspects and applications of generative AI, it's invaluable to have a consolidated reference for quick access to supplementary materials and further reading. From additional pointers to relevant datasets, to suggestions for practical projects and the latest tools, this appendix will serve as your go-to resource for expanding your knowledge and skills in generative AI.

Additional Reading

Beyond the core chapters, there are numerous publications and articles that offer further insights into specific areas of generative AI. Here are some handpicked recommendations:

- Books on machine learning and neural networks

- Research papers on the latest developments in GANs and VAEs

- Articles on ethical considerations and societal impacts of AI-generated art

Datasets and Sources for AI Art

A key component of generative AI projects is the data. The quality and relevance of your datasets can greatly influence the results. Here are some reliable sources for datasets:

- OpenAI Datasets

- UCI Machine Learning Repository

- Google Dataset Search

- Public domain art collections from museums

Tools and Libraries

While we covered some tools and libraries in Chapter 7, this list includes additional resources that can be particularly useful for various generative AI projects:

- TensorFlow: A powerful library for building and training machine learning models

- PyTorch: Known for its flexibility and ease of use in creating neural networks

- Processing: A flexible software sketchbook and language for learning how to code within the context of the visual arts

- RunwayML: A tool that makes it easy to use machine learning models in creative projects

Online Courses and Tutorials

Building on the foundations provided in this book, the following online courses and tutorials can help deepen your understanding and skillset:

- Coursera: Offers courses from top universities on AI and machine learning

- edX: Provides access to educational content and courses from institutions around the world

- Udemy: Features diverse courses that cover a wide range of topics within generative AI

- YouTube channels like Two Minute Papers and Yannic Kilcher: Perfect for keeping updated on the latest advancements and tutorials

Communities and Forums

Engaging with communities can provide support, feedback, and collaboration opportunities. Here are some popular platforms where generative AI enthusiasts and professionals converge:

- Reddit: Subreddits like r/MachineLearning, r/ArtificalIntelligence, and r/CreativeCoding

- GitHub: A great place to find open-source projects, share your code, and collaborate with others

- Kaggle: A platform for data science competitions, datasets, and community discussions

- Discord Servers: Various servers dedicated to AI and machine learning topics

Practical Projects and Templates

To put theory into practice, we've compiled a few project ideas and templates to help you get started:

- Creating AI-generated portraits using GANs - A step-by-step guide from data preparation to model training

- Generating AI music with recurrent neural networks - Templates for building a basic music generator
- Interactive AI art installations - Combining generative AI with physical interactive elements

Glossary of Terms

Refer to the glossary section for definitions and explanations of key terms and concepts used throughout this book. This will provide clarity and deepen your understanding as you explore generative AI further.

This appendix is designed to be your companion as you delve deeper into the world of generative AI. Use it as a springboard to explore new ideas, tackle ambitious projects, and join a community of like-minded innovators. The realm of generative AI is vast and brimming with possibilities—your journey has only just begun.

Glossary of Terms

This glossary aims to demystify the jargon and terminology that often accompany discussions of generative AI. As you delve deeper into the world of AI, having a clear understanding of these terms will empower you to grasp complex concepts and make the most of your creative endeavors.

Algorithm: A set of rules or instructions given to an AI system to help it learn or solve a problem.

Artificial Intelligence (AI): The simulation of human intelligence in machines that are programmed to think and learn like humans.

Autoregressive Models: Models that predict the next data point in a sequence based on the previous ones.

Backpropagation: A training algorithm for neural networks that updates the weights by propagating the error backward through the network.

Convolutional Neural Network (CNN): A type of neural network specifically designed to process and analyze grid-like data such as images.

Data Augmentation: Techniques used to increase the diversity of data available for training models without collecting new data.

Deep Learning: A subset of machine learning involving neural networks with many layers ('deep' networks) to model and understand complex patterns.

Discriminator: In GANs, the neural network that differentiates between real and synthetic data.

Generative Adversarial Network (GAN): A class of machine learning frameworks where two neural networks (a generator and a discriminator) contest with each other.

Generator: In GANs, the neural network that generates synthetic data.

Latent Space: An abstract multi-dimensional space where generative models project data during transformation processes.

Machine Learning (ML): A field of AI focused on developing systems that learn from and make decisions based on data.

Natural Language Processing (NLP): A field of AI that focuses on the interaction between computers and humans through natural language.

Neural Network: A series of algorithms that attempt to recognize underlying relationships in a set of data, designed to mimic the way the human brain operates.

Overfitting: A modeling error in machine learning where a model is too closely aligned to its training data, affecting its performance on new data.

Regularization: Techniques used to reduce overfitting by adding information or constraints to the model.

Reinforcement Learning: A type of ML where an agent learns to make decisions by taking actions in an environment to achieve maximum cumulative reward.

Transfer Learning: The process of improving learning in a new task through the transfer of knowledge from a related task that has already been learned.

Transformer: A type of model architecture designed to handle sequential data, most famous for its application in NLP tasks.

Training Data: The dataset used to train a machine learning model.

Variational Autoencoder (VAE): A type of generative model that uses principles from variational Bayesian methods and autoencoders to learn latent representations of data.

Weight: The parameters within a neural network that transform input data within the network layers, trained and adjusted during the learning process.

With this glossary, you're now equipped with a foundational understanding of key terms in generative AI. This knowledge will support your learning journey and help you navigate more complex topics as you progress.

Additional Resources

Exploring the world of Generative AI is a thrilling journey, but it can also be overwhelming due to its vast array of concepts, models, and tools. Thankfully, an abundance of resources exists to help you deepen your understanding, stay updated with the latest advancements, and connect with like-minded enthusiasts and professionals in the field. This section will guide you toward valuable materials and communities that can enrich your learning experience.

First and foremost, don't underestimate the wealth of information available in academic papers. Websites like *arXiv* (http://arxiv.org/) are invaluable for accessing cutting-edge research on neural networks, GANs, VAEs, and other generative models. These papers, authored by top researchers, provide detailed explanations, experimental results, and sometimes even source code. While they can be dense, taking the time to parse through these documents will significantly enhance your technical knowledge.

In addition to academic papers, several textbooks offer in-depth coverage of generative AI topics. Texts like "Deep Learning" by Ian Goodfellow, Yoshua Bengio, and Aaron Courville, and "Neural Networks and Deep Learning" by Michael Nielsen are staples in the AI community. These books provide foundational knowledge and advanced insights, making them essential reads for anyone serious about mastering generative AI.

Online courses and tutorials can be particularly beneficial for beginners. Platforms like *Coursera* (http://coursera.org/), *edX* (http://edx.org/), and *Udacity* (http://udacity.com/) offer courses on machine learning, neural networks, GANs, and more. Many of these courses are created by experts from top universities and tech companies, providing a structured learning path with lectures, assignments, and projects.

For hands-on practice, coding platforms like *Kaggle* (http://kaggle.com/), *Google Colab* (http://colab.research.google .com/), and *GitHub* (http://github.com/) are indispensable. *Kaggle* hosts datasets and competitions that challenge you to apply your skills in real-world scenarios. *Google Colab* offers cloud-based Jupyter notebooks, making it easy to experiment with generative AI code without worrying about setting up a local environment. *GitHub* is a treasure trove of open-source projects, libraries, and scripts that you can explore, modify, and contribute to.

Keeping abreast of the latest news and trends in generative AI is crucial as the field evolves rapidly. Blogs, podcasts, and YouTube channels run by AI enthusiasts and professionals are excellent sources of current information. *Distill* (http://distill.pub/) is a unique platform that focuses on making machine learning research more accessible through interactive and visually appealing content. You may also follow influential AI researchers and practitioners on social media platforms like Twitter and LinkedIn for updates and insights.

AI Artistry

Communities and forums provide support, advice, and networking opportunities. Websites like *Reddit* (http://reddit.com/) have vibrant subreddits such as *r/MachineLearning* and *r/ArtificialIntelligence*, where members discuss various topics and share resources. The *AI Alignment Forum* (http://alignmentforum.org/) and *EleutherAI* Discord servers are also great places to connect with others passionate about AI safety and alignment research.

Conferences and workshops offer unique opportunities to hear from world-renowned experts, witness groundbreaking work, and even present your own. Notable events include the *Conference on Neural Information Processing Systems (NeurIPS)*, the *International Conference on Learning Representations (ICLR)*, and the *Generative Modeling Summer School (GeMSS)*. Attending these events, whether physically or virtually, can provide immense learning and networking benefits.

Aside from these resources, many generative AI practitioners create their own websites or blogs where they document their projects, share tutorials, and offer guidance. Following these creators can give you insights into practical applications and creative experimentation. Python libraries like *TensorFlow* (http://tensorflow.org/) and *PyTorch* (http://pytorch.org/) have extensive documentation and community forums where you can find answers to technical questions and see how others are leveraging these tools.

Given the ethical and societal implications of generative AI, it is also wise to engage with resources that address these dimensions. The *AI Ethics Lab* (http://aiethicslab.com/) and the *Future of Life Institute* (http://futureoflife.org/) publish articles, guidelines, and research on the responsible use of AI technologies. Engaging with this content will help you understand the broader impact of your work and guide you toward ethical practices.

As you delve deeper into generative AI, you'll likely encounter challenges that require troubleshooting and optimization. When this happens, refer to comprehensive programming manuals and guides that focus on specific frameworks or models. The official documentation for libraries like *Keras* (http://keras.io/) and *Scikit-learn* (http://scikit-learn.org/) offers detailed explanations, example code, and common pitfalls, making them a go-to for solving technical issues.

Lastly, maintain a curious and exploratory mindset. The field of generative AI is rich with potential and continuously evolving. By engaging with a diverse array of resources, you'll not only deepen your expertise but also ignite innovative ideas and creative applications. The blend of theoretical knowledge, practical skills, and ethical considerations will prepare you to contribute meaningfully to the growing landscape of generative AI.

This section aims to empower you with the knowledge and tools necessary to further your journey in generative AI. Keep exploring, learning, and connecting, as the world of generative AI offers limitless possibilities for creativity and innovation.

www.ingramcontent.com/pod-product-compliance
Lightning Source LLC
Chambersburg PA
CBHW030004190526
45157CB00014B/420